Schriftenreihe der Professur für Molekulare Lebe

Band 5

Methods for the characterization of wheat flour and wheat flour dough in the context of frozen processing

Dissertation

zur Erlangung des Grades

Doktor der Ingenieurwissenschaften (Dr.-Ing.)

der Landwirtschaftlichen Fakultät
der Rheinischen Friedrich-Wilhelms-Universität Bonn

von

Julien Huen

aus
Straßburg, Frankreich

Bonn 2018

Bibliografische Information der Deutschen Nationalbibliothek
Die Deutsche Nationalbibliothek verzeichnet diese Publikation in der
Deutschen Nationalbibliografie; detaillierte bibliographische Daten sind im Internet
über http://dnb.d-nb.de abrufbar.
1. Aufl. - Göttingen: Cuvillier, 2018
    Zugl.: Bonn, Univ., Diss., 2018

Referent: Prof. Dr. Andreas Schieber
Korreferent: Prof. Dr. Peter Köhler
Tag der mündlichen Prüfung: 16. November 2018
Erscheinungsjahr: 2018

Angefertigt mit Genehmigung der Landwirtschaftlichen Fakultät der Universität Bonn

© CUVILLIER VERLAG, Göttingen 2018
    Nonnenstieg 8, 37075 Göttingen
    Telefon: 0551-54724-0
    Telefax: 0551-54724-21
    www.cuvillier.de

    ISBN 978-3-7369-9915-2
    eISBN 978-3-7369-8915-3

Table of contents

Preliminary remarks

List of abbreviations

CLSM	Confocal Laser Scanning Microscopy
DSC	Differential Scanning Calorimetry
HPLC	High-Performance Liquid Chromatography
HPAEC	High-Performance Anion-Exchange Chromatography
HPSEC	High-Pressure Size-Exclusion Chromatography
ISP	Ice Structuring Proteins
LC-MS	Liquid Chromatography – Mass Spectrometry
LV	Latent Variable
PC	Principal Component
PCA	Principal Component Analysis
PLS regression	Partial Least Squares regression
SDS-PAGE	Sodium Dodecyl Sulfate - Polyacrylamide Gel Electrophoresis
SEM	Scanning Electron Microscopy

List of publications

Huen J, Weikusat C, Bayer-Giraldi M, Weikusat I, Ringer L, Lösche, K (2014). Confocal Raman microscopy of frozen bread dough. Journal of Cereal Science, 60, 555-560.
doi: 10.1016/j.jcs.2014.07.012

Huen J, Börsmann J, Matullat I, Böhm L, Stukenborg F, Heitmann M, Zannini E, Arendt EK (2018). Wheat flour quality evaluation from the baker's perspective: comparative assessment of 18 analytical methods. European Food Research and Technology, 244, 535-545.
doi:10.1007/s00217-017-2974-3

Huen J, Börsmann J, Matullat I, Böhm L, Stukenborg F, Heitmann M, Zannini E, Arendt EK (2018).
Pilot scale investigation of the relationship between baked good properties and wheat flour analytical values. European Food Research and Technology, 244, 481-490.
doi:10.1007/s00217-017-2975 2

Conferences

Huen J, Ringer L, Lösche K, Bayer-Giraldi M (2013). The influence of a diatom antifreeze protein on ice crystallization in bread dough and its potential for improving the quality of frozen baked goods. In: Brijs K, Gebruers K, Courtin CM, Delcour JA (Eds.), C&E Spring Meeting 2013 - Unlocking the full potential of cereals: Challenges for science based innovation, Leuven, Belgium, May 29-31, 2013, p. 24.

Lösche K, Huen J (2013). Einsatz von Antifreeze-Proteinen in der Lebensmittelproduktion. FEI-Jahrestagung, Karlsruhe, Germany, 10 - 11 September 2013, https://www.fei-bonn.de/download/2013-09-11-praesentation-04-huen-loesche.pdf, accessed June 15th, 2018.

Huen J (2015). FLOURplus: Intelligent baking based on flour data. BIET'15 European meeting on baking ingredients, enzymes, and technology, Barcelona, Spain, 3 - 4 June 2015.

Huen J (2016). Erschließung der Mikrostruktur gefrorener Backwaren mittels Raman-Spektroskopie. 2. Bakers' Day, Bremerhaven, Germany, 2 - 3 March 2016.

Huen J (2016). Modellierung von Zusammenhängen zwischen Mehlanalytik, Backprozesseinstellungen sowie physikalischen und sensorischen Gebäckeigenschaften. In: Jekle M, 5. Frühjahrstagung des Weihenstephaner Instituts für Getreideforschung, Freising, Germany, 5 - 6 April 2016, p. 21.

Huen J (2016). Dynamische Rezeptur- und Prozessanpassung am Beispiel der Backbranche. In: Buckenhüskes HJ (Ed.), GDL-Kongress Lebensmitteltechnologie 2016, Lemgo, Germany, 20 - 22 October 2016.

Huen J (2016). Backpotenziale von Mehl für intelligentes Backen: Ergebnisse aus dem EU-Projekt „FLOURplus". 9. Wissenschaftliches Symposium des VDM, Würzburg, Germany, 9 - 10 November 2016.

Roeder I, Bayer-Giraldi M, Weikusat C, Huen J (2017). Influence of ISP from a polar sea-ice microalga on the microstructure of frozen cream as measured by cryo-Raman microscopy (Poster). 3rd Ice Binding Conference, Rehovot, Israel, 13 - 17 August 2017. hdl:10013/epic.51507, http://epic.awi.de/45349/1/Poster_ISP_cream_JHMB_underlined.pdf, accessed June 15th, 2018.

Huen J (2017). Einflussfaktoren der Getreideprotein-Qualität entlang der Wertschöpfungskette. 68. Tagung für Müllerei-Technologie, Detmold, Germany, 12 - 13 September 2017.

Huen J (2018). Technologies for a more efficient production and functional use of wheat protein along the value chain. Anuga FoodTec, Cologne, Germany, 20 - 23 March 2018.

Declaration of contribution as co-author

The contribution of the co-authors to the papers presented in Chapter 2, Chapter 3, and Chapter 4 are as follows:

Christian Weikusat advised on the cryo Raman microscope, supported the choice of experimental conditions, helped interpreting the Raman data and proofread the manuscript of Chapter 2.

Maddalena Bayer-Giraldi and Prof. Ilka Weikusat helped interpreting the Raman images, especially regarding ice distribution, and proofread the manuscript of Chapter 2.

Prof. Klaus Lösche proofread the manuscript of Chapter 2.

Linda Böhm born Ringer proofread all three manuscripts.

Julia Börsmann and Imke Matullat supported the experimental planning of the work presented in Chapter 2 and 3. They also proofread the corresponding manuscripts.

Mareile Heitmann performed the measurements at the Rapid Visco Analyser and the Rheo F4 reported in Chapter 3. She also proofread the manuscripts of Chapter 3 and 4.

Florian Stukenborg, Emanuele Zannini and Prof. Elke A. Arendt proofread the manuscripts of Chapter 3 and 4.

Chapter 1 – General introduction

This thesis deals with the use of deep-cooling processes in the bread supply chain and its impact on bread quality. To understand the need for deep-cooling, it is necessary to briefly describe the bread market and the related production and distribution schemes.

The worldwide bread market

Bread is a basic foodstuff in most parts of the world. The worldwide annual bread production is estimated to 125 million tons (IndexBox Marketing & Consulting, 2017). The main cereal used as a basis for bread-making is wheat, the annual worldwide production of which accounts for about 750 million tons (FAO, 2018).

In western countries, bread is nowadays available to consumers in a large variety of formats, adapted to different consumption situations. This includes fresh bread and rolls (1 to 5 days of shelf life), pre-packed long-life bread (up to 4 weeks of shelf life), and composed products for immediate to short-term consumption such as sandwiches (Fremaux, 2015). Local traditional bread types co-exist with products inspired from other countries like bagels, buns, pita, ciabatta, products based on ancient grains (ommei, spelt, quinoa, chia) and products with special nutrition claims (low-carb, protein-rich). These products are offered by a heterogeneous sales structure including traditional bakeries, supermarkets, bakery and sandwich stores, restaurants and gas stations (Eichholz-Klein, 2016).

Changes in the way of life in society in the last decades (e.g. increase in out-of-home consumption, diet trends) have not led to a decrease in bread consumption but rather to a higher variety of products and to changes in the modes of production and distribution (Fremaux, 2015). Despite high growth rates, gluten-free bakery products still represent a niche market that until now do not significantly impact the volumes of wheat-based bread (Technavio, 2017).

Modern bread-making processes – use of deep-cooling

The history of bread-making goes back to antiquity (Jacob, 2014). The basic process of bread-making involves kneading flour and water to a dough, proofing by endogenous or added microorganisms for aeration, and baking to improve digestibility related to starch gelatinisation. Although these principles have remained unchanged through the ages, the conditions of bread production have considerably evolved.

Traditionally, bread-making was performed in small businesses carrying out all production steps as well as selling at one place and in one run within a few hours (night production followed by day sale). This model is still being practiced nowadays, but its market share has considerably decreased in the last decades, while production and distribution at larger scale has gained importance: Between 2000 and 2015, the number of bakery companies in Germany has decreased from 20,302 to 12,155, while the corresponding turnover has increased from 15.7 to 19.8 billion euros (Detmers, 2017). Large scale manufacturing in factories implies that sales have to take place in separated stores, with a transportation step in-between. In addition, the manufacturing process itself may be decomposed in different segments performed at different times and places (Cauvin, 2015).

In contrast to wheat grains and wheat flour which may be stored at ambient conditions for months with only minor alteration (Fierens et al. 2015), bread dough is very unstable due to microbiological activity, and fresh bread has a limited shelf life. In this context, freezing represents a possibility to gain operational flexibility. Freezing can be performed both with bread dough and with (pre-) baked bread. Figure 1.1 gives an overview of the production process of bread with its most common variations and the points where freezing, frozen storage and thawing are typically performed.

Ingredients

| Mixing and kneading |

| batch | *or* | continuous |
(various principles & systems available)

Dough

| Forming dough pieces |

| Sheeting | *or* | Dough divider |

dough band

| cutting
folding /twisting |

Dough pieces

| Proofing |

| Direct
fermentation | *or* | Fermentation
interruption
(freezing, frozen
storage, thawing
& proofing cycle) |

leavened dough pieces

| *(pre-baking)* |

semi-finished bread

| *freezing, frozen storage, thawing* |

| final baking |

bread

| *freezing, frozen storage, thawing* |

bread

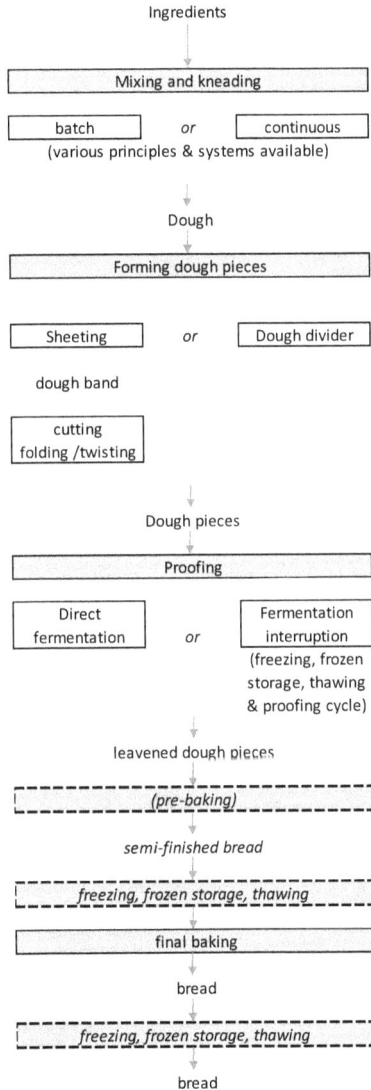

Figure 1.1: Schematic representation of the bread-making process with the different possible positions of the freezing process (based on Lösche, 2003, Cauvin, 2011 as well as own observations in European bakery companies). It should be noted that freezing is usually used once (maximum) in the production process.

In Germany, next to traditional bakeries, two main models of bread production and sales co-exist on the market: the subsidiary model and the industry model (German: "Filialisten" vs. "Zulieferbetriebe" according to the nomenclature of the German Association of Plant Bakers VDG). In the subsidiary model, bread is produced at one site and distributed once to several times a day to a network of typically 10 to 50 self-owned local points of sales (POS). The main part of the range is delivered to the POS as finished products. On top of this, a few products may be delivered as fresh dough pieces, with final baking occurring at the POS (e.g., wheat rolls and brezels). In the subsidiary model, products delivered to the POS are intended for selling on the same day. Freezing is typically used at the production site with dough pieces which are kept frozen for a few hours or a few days to increase flexibility in the production process (reducing night work, making larger dough batches, producing in advance for Sundays and bank holidays). This process is usually called fermentation interruption.

In contrast, in the industry model, bread is produced centrally and intended for national or international distribution in the frozen state (with the exception of pre-packed long-life bread), mainly to other companies (supermarkets, bakery and sandwich stores, restaurants, gas stations). In this case, frozen storage is typically performed for several months, with shelf lives of around one year. In this model, there is a high need to keep the handling at the point of sale as easy as possible, as the sales staff is seldom specialized in baking. While in earlier times, this model was often practiced with frozen dough pieces that needed to be thawed, proofed and baked at the POS, the main current practice involves (part)-baking the bread in the factory and freezing after baking. In this way, at the POS, only thawing and final baking are necessary. In the case of products with soft crust like bagels, a final baking is not even necessary, which further simplifies operations.

The subsidiary model is characterised by relatively small batches, as dough pieces are produced only a few days in advance. The production processes are typically of the batch or semi-batch type, with manual handling between the single steps as well as partly manual work in the single processes.

In contrast, the industry model typically involves continuous processes at high throughput (several tons per hour on each production line, corresponding to several thousand pieces per hour). As production is entirely disconnected from sale and consumption in time, very large batches can be produced.

Of course, the boundaries between the subsidiary and the industry model are open and in the single companies, different variations or combinations of these models may be practiced. For instance, companies operating primarily according to the subsidiary model may choose to integrate some industry products in their range, enabling them to offer a higher diversity of products.

As described above, freezing has become a major process in the bread production and distribution chain. Although low temperatures stop microbiological growth and considerably delay chemical reactions, the processes of freezing, frozen storage and thawing often lead to quality deterioration. In order to gain a better understanding, it is necessary to describe the composition of flour and bread dough, the main quality attributes of bread products, and the physical processes occurring during freezing and thawing.

Wheat flour composition and characterization

The wheat flour supply chain is mainly composed of wheat breeders, farmers, cereal traders and millers. All members of this supply chain have an influence on the composition of the flour finally delivered to bakery companies. Each wheat variety as developed by the breeders and approved by the responsible national authorities (in Germany: Bundessortenamt) has a certain genetic potential to synthetize specific grain components (Henry & Wrigley, 2018). The growing conditions (soil, climate, fertilisation) have a major effect on gene expression and thus on the grain composition (Koga et al, 2016, Hawkesford et al, 2014). After harvest, single grain batches are stored together in silo cells. In most cases, this occurs at the facilities of grain trading companies (VDM, 2016). The decision about which grain batches will be stored together directly impacts the quality that is delivered to the mill. The mill, in turn, influences the flour composition through (i) blending different grain batches, (ii) setting the milling parameters, (iii) combining milling fractions in a specific way (Brütsch, 2017) and (iv) including additives like ascorbic acid and malt flour (common in Germany) as well as enzymes (common e.g. in France).

Wheat flour is mainly composed of starch (70 - 75 %), water ($\approx$ 14 %), proteins (10 - 13 %), non-starch polysaccharides (arabinoxylan 2 - 3 %), fat ($\approx$ 2 %), endogenous enzymes ($\approx$ 1 %) and minerals ($\approx$ 0.5 % depending on the selected extraction rate) (Goesaert et al., 2005).

Analytical investigations usually conducted on flour include the quantification of single components and the description of functionality, especially based on rheology. Methods appropriate for measurements in the processing companies are described in the ICC and AACC standards. Measurements performed in mills typically include water content, minerals (ash), total proteins, gluten (dry and wet), falling number, sedimentation value and damaged starch (ICC 104/1, 105/2, 107/1, 110/1, 116/1, 155, 172). In addition, dough kneading and stretching properties, gluten aggregation properties and starch gelatinization may be assessed by specific instruments and methods (ICC 114/1, 115/1, 121, 126/1, 173, AACC 56-11). Cereal traders and mills also use NIR spectroscopy for the fast determination of humidity and protein in grains at delivery.

In research, more sophisticated techniques are used for quantifying single components, including HPLC, SDS-PAGE and LC-MS for gluten analysis (Wrigley, 1996, Wrigley et al., 2006, Schalk, 2017), HPSEC, HPAEC and X-ray diffraction for starch analysis (Grant et al., 2002, Yoo et al., 2002, Jane et al. 1999) and HPSEC and HPAEC for non-starch polysaccharides (Courtin and Delcour, 2002, Ordaz-Ortiz, 2005).

From flour to bread dough

Next to flour, the main ingredients of bread dough are water, yeast or sourdough and salt. Optionally, sugar, fat, malt extracts, vital gluten, ascorbic acid, emulsifiers, acidity regulators, hydrocolloids and enzymes may be added to improve processability and/or baked good quality (Wassermann, 2009). It should be noted that, while rationalization of bread-making has first led to an increased use of additives and enzymes in bread-making, the market now increasingly demands "clean label" products (Fremaux, 2015). The satisfaction of this demand in large-scale production implies a better process expertise and the use of flours with improved functionality.

Dough kneading is essentially characterised by the homogenisation of dough components and the hydration of starch granules (especially damaged starch), gluten and non-starch polysaccharides. In addition under mechanical shear, a 3-dimensional gluten network is formed. During network formation, new disulfide bonds are being created by the oxidation of S-H groups, existing disulfide bonds are being exchanged among each other (so-called disulfide interchange reactions), and hydrogen bonds are being formed between different parts of the glutenin polymer as well as with gliadins. The glutenin polymer is believed to be responsible for the elastic properties of the dough, while gliadins and starch increase viscosity (Wrigley et al., 2006).

Quality attributes of bread

Bread products are characterised by a number of physical and sensory attributes. The main physical attributes include weight, specific volume, dimensions, colour of crust and crumb, pore size and distribution in the crumb, hardness and elasticity of crumb (Scanlon and Zghal, 2001). Sensory attributes include appearance, smell, taste and texture/mouthfeel, for each of which a number of descriptors are used (Callejo, 2011).

In b2b (business to business) trading, a detailed specification is issued for each product. In these specifications, typically only a narrow corridor is defined for the main attributes. Therefore, a high degree of reproducibility is required in industrial production.

In traditional bakeries and in subsidiary bakeries, which operate in the b2c (business to consumer) model, no specification is agreed with the customer. Therefore, a higher degree of variability is possible. Nevertheless, these companies must offer a level of organoleptic quality comparable or superior to industrial companies, as they are in direct competition with them and have higher per-piece production costs.

Conditions of freezing, frozen storage and thawing

Two scenarios have to be differentiated here. In the case of fermentation interruption, freezing, frozen storage, thawing and proofing may be performed in a single fermentation interruption cabinet according to a programmed temperature and humidity profile if the storage time is only a few hours. Typical freezing temperatures are in the range of -20 °C to -10 °C. In contrast, if the dough pieces have to be stored

17

for several days, freezing will occur in a shock freezer at -40 to -30 °C, storage in a cold room at -20 to -18°C, followed by controlled thawing and proofing in a fermentation interruption cabinet (Lösche, 2003).

In industry production, freezing of bread typically takes place in continuous freezing tunnels operating in with air temperatures of -40 °C to -30 °C. Storage and transportation occur at -20 to -18°C. Thawing at the point of sale occurs at room temperature without particular equipment (Lösche, 2003).

Phenomena occurring in bread dough during freezing, frozen storage and thawing

Scientific literature and experience reported by bakery companies describe negative effects related to the use of freezing with bread dough. These are mainly a reduction in volume, a coarser crumb structure and a loss of dough firmness leading to flatter products (Gélinas et al., 1996, Lu and Grant , 1999, Esselink et al, 2003, Frauenlob et al., 2017). So far, the underlying phenomena are only partially understood.

The most obvious effect when freezing bread dough is the transformation of liquid water into ice. In bread dough, some water molecules are involved in the hydration of macromolecules (gluten, starch or non-starch polysaccharides) by hydrogen bonds. Another part of the water molecules acts as a solvent for water-soluble substances like mono- and disaccharides and minerals.

Ice formation may be monitored quantitatively by differential scanning calorimetry (DSC). DSC measurements show that, when decreasing the temperature of bread dough, ice formation does not start at 0 °C but at temperatures around -5 °C (the exact temperature depending on the formulation). This can be related to the substances dissolved in water. When decreasing temperature further, more ice is created, while a liquid water phase remains. Because ice is composed only of water molecules, the concentration of solutes increases in the water phase as the proportion of ice rises. At even deeper temperatures (-44°C in the experiments of Baier-Schenk et al., 2005a and 2005b), the remaining liquid water phase undergoes a glass transition, visible in the DSC through a change in heat capacity. The glassy state is an amorphous solid state.

Next to this, there is a glass transition of gluten, occurring at around -13 °C in the dough (Kalichevsky et al., 1992, Noel et al., 1995, Baier-Schenk et al., 2005a).

Ice has a crystalline structure. Crystallization is an exothermic process, as ice has a lower internal energy than liquid water. Ice formation implies the creation of crystal nuclei. A high activation energy is necessary for homogeneous nucleation (Koop et al., 2000). Once a nucleus is present, further water molecules can join the existing crystalline structure with a relatively lower activation energy (crystal growth). Experiments on pure water and on aqueous solutions show that in the case of high temperature gradients, leading to a high undercooling, nucleation is facilitated. This favours the creation of a high number of small crystals. In contrast, in the case of a low temperature gradient, leading to a low undercooling, a limited number of larger crystals are created (Petzold and Aguilera, 2009). During frozen storage, ice crystals may undergo transformations. This recrystallization leads to a smaller number of larger crystals, which are thermodynamically more stable than small crystals. Recrystallization involves migration of water molecules between crystals. There are several mechanisms for this, depending on whether the transfer of water molecules occurs via the liquid, the gas or the solid phase. Temperature variations during storage favours recrystallization through the liquid phase. In addition, recrystallization is temperature-dependant and is slower at lower temperatures. During thawing, recrystallization may occur as temperature increases, leading to the short-time formation of larger crystals, before these finally melt (Petzold and Aguilera, 2009).

In the case of pure water and aqueous solutions, these phenomena can well be monitored, as crystal boundaries can be observed via light microscopy. Freezing experiments on vegetal and animal tissue (in the case of food: fruit, vegetables, fish and meat) show that the phenomena described above also apply to cellular systems. In addition, it was shown that larger crystal sizes are responsible for higher cell damages (Martino et al., 1998, Delgado et al., 2005, Otero et al., 2000, Sun, 2003, Do et al., 2004). This justifies the use of shock freezers operating at low temperatures, in order to create high temperature gradients and small crystals.

In bread dough, ice crystals are difficult to observe, and there are only few publications showing ice crystals either directly or indirectly. Baier-Schenk (2005a) published pictures obtained by cryo SEM (scanning electron microscopy), showing ice crystals of several hundred micrometres in diameter in the pores of the dough after 197 days of storage at -22°C. These crystals had a very regular shape with a clearly identifiable hexagonal basal face and prismatic faces. Such large crystals were not found in the pores of freshly frozen dough, evidently showing a recrystallization effect over time. This observation, however, does not provide any information on the structure of ice contained in the dough matrix itself. In pores, ice crystals can grow without any mechanical constraints, which can explain the regular shape observed.

In the case of vegetal and animal tissues and of microbial cells, freezing damage is mainly explained by damage to cell components, especially cell membranes, due to ice crystals. In bread dough, the observed quality losses are believed to the effect of ice crystals on yeast cells and on gluten. The latter, however, is more based on the fact that gluten is responsible for dough strength and gas holding capacity than on tangible knowledge on the interaction of gluten with ice.

Reduction of deep-cooling damage in bakery products

Developing strategies for reducing freezing damage implies understanding the underlying phenomena. From a practical point of view, the options open to the baker are to improve either the recipe or the production process. On the recipe side, specific ingredients may be added. These include substances that decrease the freezing point and substances that increase the temperature of glass transition (Levine and Slade, 1990). Bakery improvers combining several of these substances, especially mono- and disaccharides as well as the hydrocolloids guar gum and carboxymethylcellulose CMC are available on the market with the specific claim of improving fermentation interruption performance.

On the process side, one may reduce the freezing temperature, reduce frozen storage temperature, avoid storage temperature variations, and improve thawing conditions. Shock freezing is commonly used with bread dough and bread without evidence of its impact on crystal structure.

Interestingly, so far, strategies on the ingredient size mainly consist in adding extra substances to the dough. Little is known about flour qualities more appropriate for freezing processes and no flours are marketed with the specific claim of being particularly suitable for this purpose.

Raman spectroscopy and its use in confocal microscopy

Raman spectroscopy consists of measuring the shifting of wavelength occurring in light as a consequence of its interaction with molecules during inelastic scattering. Raman spectra have been described in literature for a great variety of substances, including food components. The integration of a Raman spectrometer in confocal microscope allows measuring a Raman spectrum for each point of the sample on which the laser light is focused. If the composition of the sample is known and reference Raman spectra are available for each of these substances, it is possible to determine the relative concentration of each substance at each point of the sample investigated. By scanning a portion of the sample in two or three dimensions, images showing the special distribution of each substance within the investigated area or volume can be created, as well as multi-phase images showing the respective distribution of several substances.

This technique was developed in the late 1990s and has found applications in a variety of research areas, like pharmacy, mineralogy, petrography, polymer science and glaciology (Toporski et al., 2018, Krafft et al., 2012, Weikusat et al., 2012). Its main disadvantage is the possibility to identify and image substances without needing a specific staining. Its main inconvenient is the weakness of the Raman signal. As a result, relatively long integration times are necessary. In contrast, in classical Confocal Laser Scanning Microscopy, the sample is stained with one or several fluorescent substances. Due to the high intensity of the fluorescence signal, fast measurements are possible.

In food research, the use of Raman microscopy is only emerging, with first applications on fat spreads, bovine milk, mayonnaise, cheese and soy drinks (Dalen et al., 2017, Gallier et al., 2011, Roeffaers et al., 2011). For investigating frozen food products, Raman microscopy seems to be particularly appropriate, as liquid water and ice have different Raman spectra, due to the different state of the O-H binding. Staining of ice is not possible, as ice crystals consist only of water molecules.

Aims of the thesis

This thesis aims to contribute to solving problems linked to quality damages due to the use of freezing in bakery companies. This contribution is targeted at two levels:

- At the level of fundamental research, the aim is to develop an imaging method appropriate for investigating the microstructure of frozen dough. This method should be especially suitable to show the microstructure of ice in relation to the other dough components. This is intended to open the way for further investigations of crystallization and recrystallization phenomena in bread dough as well as the interactions of ice with the other dough components, especially gluten, providing a better understanding of the phenomena leading to quality losses through deep-cooling.
This topic is addressed in Chapter 2.

- At the level of applied research, the aim is to identify among the analytical methods usually used within the wheat supply chain those which are most appropriate to give indications on the suitability of flours for freezing processes. The corresponding results are intended to help bakers and millers specifying flours suitable for dough freezing. They may also be of help for the upstream stakeholders of the wheat supply chain (wheat breeders, farmers and cereal traders), who could adapt their processes for delivering the required wheat quality in a targeted way. This topic is addressed in Chapters 3 and 4, which describe two parts of a single broad experimental setup and dataset. In Chapter 3, the most current analytical methods are compared, before they are brought in relationship to bakery performance, especially in the case of fermentation interruption, in Chapter 4.

References

Baier-Schenk A, Handschin S, Conde-Petit B (2005a). Ice in prefermented frozen bread dough—an investigation based on calorimetry and microscopy. Cereal Chemistry 82, 251-255.

Baier-Schenk A, Handschin S, von Schönau M, Bittermann AG, Bächi T, Conde-Petit B (2005b). In situ observation of the freezing process in wheat dough by confocal laser scanning microscopy (CLSM): formation of ice and changes in the gluten network. Journal of Cereal Science 42, 255-260.

Brütsch L, Huggler, I, Kuster S, Windhab EJ (2017). Industrial roller milling process characterisation for targeted bread quality optimization. Food Bioprocess and Technology 10, 710-719.

Callejo MJ (2011). Present situation on the descriptive sensory analysis of bread. Journal of Sensory Studies 26, 255-268.

Cauvain S (2015). Technology of Breadmaking, 3rd edition, Springer.

Courtin CM, Delcour JA (2002). Arabinoxylans and endoxylanases in wheat flour bread-making. Journal of Cereal Science 35, 225-243.

Dalen G, Velzen EJJ, Heussen PCM, Sovago M, Malssen KF, Duynhoven JPM (2017). Raman hyperspectral imaging and analysis of fat spreads. Journal of Raman Spectroscopy 48, 1075-1084.

Delgado AE, Rubiolo AC (2005). Microstructural changes in strawberry after freezing and thawing processes. LWT - Food Science and Technology 38, 135-142.

Detmers U (2017). Pressekonferenz des Verbands Deutscher Großbäckereien e.V., 20.09.2017, http://grossbaecker.de/pressemeldungen-detail/statement-prof-dr-ulrike-detmers.html, accessed February 4th, 2018.

Do GS, Sagara Y, Tabata M, Kudoh KI, Higuchi T (2004). Three-dimensional measurement of ice crystals in frozen beef with a micro-slicer image processing system. International Journal of Refrigeration 27, 184-190.

Eichholz-Klein S (2016). Das Backgewerbe 2016. Institut für Handelsforschung, Köln.

Esselink FJ, van Aalst H, Maliepaard M, van Duynhoven PM (2003) Long-term storage effect in frozen dough by spectroscopy and microscopy. Cereal Chemistry 80, 396-403.

Fierens E, Helsmoortel L, Joye IJ, Courtin CM, Delcour JA (2015) Changes in wheat (*Triticum aestivum L.*) flour pasting characteristics as a result of storage and their underlying mechanisms. Journal of Cereal Science 65, 81–87.

Food and Agriculture Organisation of the United Nations (2018), FAO Cereal Supply and Demand Brief, http://www.fao.org/worldfoodsituation/csdb/en, accessed February 2nd, 2018.

Frauenlob J, Moriano ME, Innerkofler U, D'Amico S, Lucisano M, Schoenlechner R (2017). Effect of physicochemical and empirical rheological wheat flour properties on quality parameters of bread made from pre-fermented frozen dough. Journal of Cereal Science 77, 58-65.

Fremaux A (2015). Bake-off bakery markets in the EU. Gira, Paris.

Gallier S, Gordon KC, Jiménez-Flores R, Everett, DW (2011). Composition of bovine milk fat globules by confocal Raman microscopy. International Dairy Journal 21, 402-412.

Gélinas P, McKinnon CM, Lukow OM, Townley-Smith F (1996). Rapid evaluation of frozen and fresh dough involving stress conditions. Cereal Chemistry 73, 767–769.

Goesaert H, Brijs K, Veraverbeke WS, Courtin CM, Gebruers K, Delcour JA (2005). Wheat flour constituents: how they impact bread quality, and how to impact their functionality. Trends in Food Science and Technology 16, 12 30.

Grant LA, Ostenson AM, Rayas-Duarte P (2002). Determination of Amylose and Amylopectin of Wheat Starch Using High Performance Size-Exclusion Chromatography (HPSEC). Cereal Chemistry 79, 771-773.

Hawkesford MJ (2014). Reducing the reliance on nitrogen fertilizer for wheat production. Journal of Cereal Science 59, 276-283.

Henry RJ, Wrigley CW (2018). Towards a genetic road map of wheat-processing quality. Journal of Cereal Science 79, 516-517.

IndexBox Marketing & Consulting (2017). World: Bread and bakery product - Market report. Analysis and forecast to 2025. Douglas, United Kingdom.

Jacob HE (2014). Six thousand years of bread: Its holy and unholy history. Skyhorse Publishing, New York.

Jane J, Chen YY, Lee LF, Mcpherson AE, Wong KS, Radosavljevic M, Kasemsuwan T (1999). Effects of amylopectin branch chain length and amylose content on the gelatinization and pasting properties of starch. Cereal Chemistry 76, 629-637.

Kalichevsky MT, Jaroszkiewicz EM, Blanshard JMV (1992). Glass transition of gluten. 1: Gluten and gluten-sugar mixtures. International Journal of Biological Macromolecules 14, 257-266.

Koga S, Böcker U, Uhlen AK, Hoel B, Moldestad A (2016). Investigating environmental factors that cause extreme gluten quality deficiency in winter wheat (*Triticum aestivum L.*). Acta Agriculturae Scandinavica, Section B – Soil & Plant Science 66, 237-246.

Koop T, Luo B, Tsias A, Peter T (2000). Water activity as the determinant for homogeneous ice nucleation in aqueous solutions. Nature 406, 611.

Krafft C, Dietzek B, Schmitt M, Popp J (2012). Raman and coherent anti-Stokes Raman scattering microspectroscopy for biomedical applications. Journal of Biomedical Optics 17, 040801.

Levine H, Slade L (1990). Cryostabilization technology: theromoanalytical evaluation of food ingredients and systems. In: Harwalkar VR, MA CY, (Eds.), Thermal analysis of foods. Elsevier Applied Science, London, 221-305.

Lösche K (ed.) (2013). Kältetechnologie in der Bäckerei. Behr's Verlag, Hamburg.

Lu W, Grant LA (1999). Effects of prolonged storage at freezing temperatures on starch and baking quality of frozen doughs. Cereal Chemistry 76, 656-662.

Martino MN, Otero L, Sanz PD, Zaritzky NE (1998). Size and location of ice crystals in pork frozen by high-pressure-assisted freezing as compared to classical methods. Meat Science 50, 303-313.

Noel TR, Parker R, Ring SG, Tatham AS (1995). The glass-transition behaviour of wheat gluten proteins. International Journal of Biological Macromolecules 17, 81-85.

Ordaz-Ortiz JJ, Saulnier L (2005). Structural variability of arabinoxylans from wheat flour. Comparison of water-extractable and xylanase-extractable arabinoxylans. Journal of Cereal Science 42, 119-125.

Otero L, Martino M, Zaritzky N, Solas M, Sanz PD (2000). Preservation of microstructure in peach and mango during high-pressure-shift freezing. Journal of Food Science 65, 466-470.

Petzold G, Aguilera JM (2009). Ice morphology: fundamentals and technological applications in foods. Food Biophysics 4, 378-396.

Roeffaers MB, Zhang X, Freudiger CW, Saar BG, Xie XS, van Ruijven M, Dalen G, Xiao C (2011). Label-free imaging of biomolecules in food products using stimulated Raman microscopy. Journal of Biomedical Optics, 16, 021118.

Scanlon MG, Zghal MC (2001). Bread properties and crumb structure. Food Research International, 34, 841-864.

Schalk K, Lexhaller B, Koehler P, Scherf KA (2017). Isolation and characterization of gluten protein types from wheat, rye, barley and oats for use as reference materials. PLoS ONE 12: e0172810.

Sun DW, Li B (2003). Microstructural change of potato tissues frozen by ultrasound-assisted immersion freezing. Journal of Food Engineering 57, 337-345.

Technavio (2017). Global Gluten-Free Bakery Market 2017-2021, Report ID: 5126537.

Toporski J, Dieing T, Hollricher O (2018). Confocal Raman Microscopy. Springer, Heidelberg.

VDM, Verband Deutscher Mühlen e.V. (2016). Die Bedeutung der Mühlenwirtschaft in der Wertschöpfungskette, http://www.muehlen.org/fileadmin/Dateien/8_Presse_Service/ 2_Fotos_Infografiken/VDM_Grafik_Bedeutung_der_Muehlenwirtschaft_2016.pdf, accessed February 4th, 2018.

Weikusat C, Freitag J, Kipfstuhl S (2012). Raman spectroscopy of gaseous inclusions in EDML ice core: first results – microbubbles. Journal of Glaciology 58, 761-766.

Wrigley CW (1996). Giant proteins with flour power. Nature 381, 738-739.

Wrigley CW, Békés F, Bushuk W (2006). Gluten: A balance of gliadin and glutenin. Gliadin and glutenin. The unique balance of wheat quality. AACC International Press, St Paul, 3-32.

Chapter 2 – Confocal Raman microscopy of frozen bread dough

Abstract

The use of freezing technology is well established in industrial and craft bakeries and is still gaining importance. In order to optimize recipes and processes of frozen baked goods, it is essential to be able to investigate the products' microstructure. Especially ice crystals and their interaction with the other components of the frozen products are of interest. In this study, frozen wheat bread dough was investigated by confocal Raman microscopy. The Raman spectra measured within the dough were compared with spectra of the main components of frozen dough, i.e. ice, liquid water, starch, gluten and yeast. In this way, the spatial distribution of the single components within the dough was determined and corresponding images of the frozen dough microstructure were generated. On these images, ice appears as a continuous network rather than as isolated crystals. We suggest that this method may be appropriate for characterizing crystallization phenomena in frozen baked goods, allowing to better understand the reasons for quality losses and to develop strategies for avoiding such losses.

Keywords

Confocal Raman microscopy, Frozen bread dough, Microstructure, Ice crystals

---

This chapter has been published:

Huen J, Weikusat C, Bayer-Giraldi M, Weikusat I, Ringer L, Lösche, K (2014). Confocal Raman microscopy of frozen bread dough. *Journal of Cereal Science*, 60 (3), 555-560. doi: 10.1016/j.jcs.2014.07.012

Introduction

As the bakery business is being concentrated and rationalized, increasing use is made of freezing technology in production and distribution (Le Bail et al., 2012). Freezing allows a separation in time and space of process operations that would traditionally be performed in one run and in one place.

In bread-making, freezing is used at several stages of production: for non-fermented or partly-fermented dough, for partly or fully baked products (Le Bail et al., 2012). Depending on the application, the products are kept frozen for a few hours or for several weeks or months. A large variety of equipment, including shock-freezers, fermentation interrupters, climatic chambers, and cold storage rooms are used for realizing the operations of freezing, cold storage and thawing.

Although the intention when using freezing is to keep the product in a steady state, in practice a number of physical and chemical phenomena occur, affecting the quality of the final product in a mostly negative way. Among these phenomena, the formation of ice crystals is believed to be of primary importance for two main reasons (Berglund et al.,1991; Baier-Schenk et al., 2005a): (1) Ice crystals are made of pure water which is being separated from the product matrix. Cryoconcentration occurs in the liquid phase, which may influence the solubility of proteins and the activity of enzymes. During storage, ice crystals grow due to recrystallization, especially in the pores, thus further modifying the distribution of water in the product. (2) Ice crystals may mechanically damage the dough components, especially the gluten network and the yeast cells, because the freezing front exerts stress on the surrounding material. This effect is believed to be more pronounced as the crystal size increases due to recrystallization.

In order to optimize the recipes and the production processes of frozen baked goods, it is essential to be able to monitor the phenomena occurring in the products in the frozen state. Differential scanning calorimetry (DSC) allows quantitative investigations of ice crystallization. For monitoring the size and the distribution of the ice crystals as well as their mechanical interactions with the other components of the dough, imaging techniques are required. So far, scanning electron microscopy in the frozen state (cryo SEM, Zounis et al., 2002, Esselink et al., 2003; Baier-Schenk et al., 2005a) and confocal laser scanning microscopy (CLSM, Baier-Schenk et al., 2005b) have been used for that purpose. Cryo SEM has allowed demonstrating the growth of ice crystals within the pores over storage time and CLSM to identify regions of preferential nucleation. However, in both techniques, a difficulty is the limited possibility to unambiguously differentiate the ice crystals from the other components of the dough. In cryo SEM, this differentiation is performed based on the regular shape of the crystals – but this is only valid in the pores, where ice crystals can grow without spatial constraints. In CLSM, changes in the reflection properties were attributed to ice crystal growth. However, this method did not allow for generating precise images of the ice crystal structure. Due to these limitations, little is known about the structure of the ice crystals that are entrapped in the dough matrix, which yet represent the main part of the frozen water.

Raman spectroscopy belongs to the group of vibrational spectroscopies (Smith and Dent, 2005). It utilizes the inelastic scattering of light photons on molecules or molecular groups, called Raman effect. If the molecule (or group) has suitable vibrational modes, a photon can transfer a fraction of its energy to the vibration (Stokes scattering). The positions of the Raman bands directly give the energy of the detected vibrations. The ensemble of Raman active vibrations is characteristic for each compound and can range from single bands to very complex multi-band spectra. Raman spectroscopy is a non-destructive method requiring very little sample preparation and it is suitable for a wide range of materials. If high quality reference spectra are available, it is a very sensitive tool for phase identification.

With the implementation of Raman spectroscopy in confocal microscopy in the late 1990s, it became possible to use Raman data for microimaging purposes. Applications were developed in a variety of scientific fields including mineralogy, petrography, polymer science, pharmaceutical research (Dieing et al., 2011), biomedical diagnostics (Krafft et al., 2012) and glaciology (Weikusat et al., 2012). In agricultural and food science and more specifically in cereal science, only little use has been made of this technique so far. Piot et al. (2000, 2001, 2002) used confocal Raman microscopy for exploring the spatial distribution of starch, gluten, arabinoxylan and ferulic acid in wheat grains. Recently, Jääskeläinen et al. (2013) performed similar investigations with higher (sub-µm) spatial resolution on barley and wheat grains.

Based on the fact that confocal Raman microscopy has shown to be suitable for characterising both ice crystals and the main components of cereals, our objective was to develop a measurement method appropriate for investigating the microstructure of frozen bread dough.

Experimental

Raw materials and equipment

The following ingredients were used in the experiments: Wheat flour type 550 (Roland Mühle, Germany), compressed yeast (Frischhefe, Deutsche Hefewerke GmbH, Germany), and salt (Suprasel fine, Suprasel, The Netherlands).

Raman measurements were performed on a WITec Alpha 300R microspectroscopy system equipped with a frequency-doubled Nd:YAG laser (λ = 532 nm), an UHTS300 Raman Spectrometer (grating: 600 grooves/mm, pixel resolution <0.09 nm) with a Peltier-cooled DV401A-BV CCD detector (peak quantum efficiency at ~550 nm and -60°C: >95%) and a 50x LWD objective, operated in a cold laboratory at -15 °C at the Alfred-Wegener Institute. The laser power on the sample was <30 mW.

Assessment of Raman spectra of single dough components

The Raman spectra of ice, liquid water, starch, gluten, and yeast were assessed using the following procedure.

Sample preparation

A 3.5% (w/v) salt solution in bidistilled water was prepared. One droplet of this solution (20 mL) was placed on a microscope slide, covered with a cover slip using a 2 mm spacer to standardize thickness, and frozen at -20 °C. In this way ice crystals and a liquid phase (cryoconcentrated salt solution) were formed. The salt present in the liquid phase is expected to influence the Raman spectrum only to a minimal extent, as its main component NaCl (≥99.8% according to the supplier's specification) has no molecular vibration.

Wheat flour was hydrated and separated into a starch suspension and a wet gluten piece using a Glutomatic 2200 from Perten Instruments, Sweden. One droplet of the starch suspension was placed on a microscope slide, covered with a cover slip using a 2 mm spacer and frozen at -20 °C. The same was done with a small portion of the wet gluten piece and of the compressed yeast block.

Measurement

The Raman spectrum of each of the samples representative for the individual dough components was measured at 10 different points, with 20 accumulations of 1 s each per point, and the average spectrum was calculated for each component.

Dough sample preparation

Three frozen dough samples were prepared at three different days in the following way: 50 g of wheat flour, 28 g of bidistilled water, 1.5 g of compressed yeast and 1 g of salt were mixed and kneaded to a dough in a Brabender Farinograph AT at 20 °C. The mixing time was 2 min at 36 rpm and the kneading time 4 min at 63 rpm. After kneading, the dough was allowed to rest for 15 min at room temperature. Subsequently, a small piece (approx. 250 mg) of the inner part of the dough was cut out, placed on a microscope slide, covered with a cover slip using a 2 mm spacer and frozen at -20 °C.

Confocal Raman microscopy of frozen dough samples

On the day following preparation, the samples were transferred to the microscopy laboratory at -15 °C. Before measurement, the samples were kept for at least one hour at -15 °C to stabilize at that temperature.

For each of the 3 frozen dough samples, an area of 100 x 100 µm was measured with a resolution of 200 x 200 points and an integration time of 1 s per point, resulting in a measurement time of approx. 12 h.

Confocal Raman microscopy: data processing and imaging

The data from the area scans were processed in two different ways to produce images showing the spatial distribution of the single dough components (ice, liquid water, starch, gluten and yeast).

In the first method, single Raman bands characteristic for each component were integrated. Monochrome images were generated representing the intensity of the individual bands at each measurement point. The spectral ranges of the chosen bands are given in Table 2.1 and are marked in blue in Fig. 2.1.

The second method considered the full Raman spectra instead of single bands. In that method, the Raman spectrum measured at each point of the sample was assumed to be a linear combination of the spectra of the single dough components. After performing a 3rd order polynomial background subtraction on all spectra, a multiple linear regression was completed using the function Basis Analysis of the WITec Project software (release 2.10, WITec GmbH, Ulm, Germany). The assessed regression coefficients were used as indicators of the concentration of the individual dough components, and corresponding monochrome images were generated. This method is well established in confocal Raman microscopy and was used among others by Jääskeläinen et al. (2013).

The monochromatic images showing the distribution of the single components were combined to colour images in which each colour represents one component. This allows visualizing the position and distribution of the components relative to each other.

Table 2.1: Spectral ranges selected as characteristic for the single dough components. The references cited provide a detailed discussion of the Raman spectra and of the band assignment for the individual dough components.

Dough component	Spectral range (cm^{-1})	Band assignment	Reference
Ice	3080-3200	OH stretching band	Đuričković et al., 2011
Starch	460-510	Stretching vibration of the carbon network	Piot et al., 2000 Fechner et al., 2005
Gluten	1645-1690	Amide I (partly)	Piot et al., 2000
Yeast	740-766	Ring breathing vibration of tryptophan (possibly)	Rösch et al., 2006

Results

Raman spectra of single dough components

The measured spectra of the single dough components are presented in Fig. 2.1, and bands of particular interest are listed in Table 2.1.

Ice is especially characterized by the OH stretching band with a maximum in the spectral range of 3080-3200 cm^{-1}, as described by Đuričković et al. (2011). This band was not found in the spectra of the other dough components. The spectrum of liquid water is dominated by the OH stretching band with a maximum in the range of 3300-3420 cm^{-1}, and also embodies the OH bending band (1580-1640 cm^{-1}). Starch shows a series of bands, reflecting its molecular complexity. These bands were already reported by Piot et al. (2000) and Fechner et al. (2005) and their assignment discussed by these authors. The CH stretching band (2800-3050 cm^{-1}) and the OH stretching band are both strongly represented. The narrow band in the range of 460-510 cm^{-1}, which is attributed to the stretching vibration of the carbon network of starch, was not found in spectra of the other dough components.

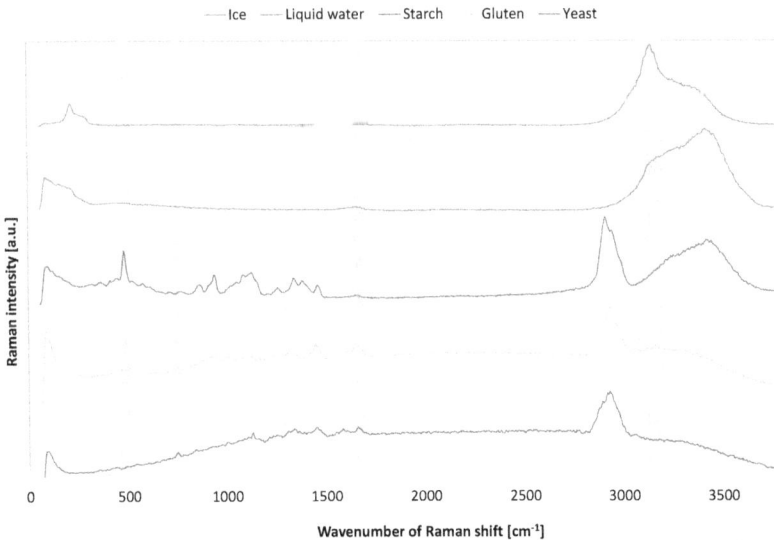

Figure 2.1: Raman spectra of single dough components: ice, liquid water, starch, gluten and yeast. The spectral ranges used for band integration imaging are marked in blue.

The gluten spectrum shows a higher base signal, due to fluorescence, and a series of bands that were described and which assignment was discussed by Piot et al. (2000). As in starch, the CH stretching band is strongly represented. The band in the range of 1645-1690 cm^{-1} is attributable, at least partly, to amide I (see "band position and assignment" in the discussion).

Yeast also shows a high base signal caused by fluorescence. The observed bands are in line with the measurements of Rösch and Harz (2006) with Saccharomyces cerevisiae. In the range of 1645-1690 cm^{-1}, yeast show a signal similar to gluten, yet with lower intensity. The band in the range of 740-766 cm^{-1}, which was assigned by the latter authors to tryptophan, appears to be specific for yeast in the studied dough system in terms of intensity – gluten and to a lesser extent starch also have bands in this spectral range, but they are weaker.

Raman images of frozen dough

The Figs. 2.2-2.4 show the spatial distribution of the individual dough components within the three samples, as determined by both data processing methods (band integration and multiple linear regression). The distribution of liquid water, however, was determined only by multiple linear regression, as the OH Raman bands of liquid water overlap with the OH bands of the starch and the gluten spectra – in other words, liquid water has no specific single Raman band in the frozen dough system. Multiple linear regression, on the other hand, failed to allow for a determination of the distribution of yeast, as is discussed below. For each sample, a colour image showing the relative spatial distribution of the single dough components was generated.

Both representations are complementary. The monochrome images give more details about the structure of the single components. Due to the depth of field of a few µm, they give some insights into the 3-dimensional structure. Elements located a few µm above or below the focal plane are still being detected but lead to a weaker signal, which is represented by a lower pixel brightness. The colour images, on the other hand, show how the dough components are spatially organized relative to each other in the focal plane.

Starch appears on the pictures as large granules with a diameter of 20-25 µm and smaller granules with a diameter of 2-5 µm. Gluten appears as fibrils organised around the starch granules, partly with a spatial orientation as parallel strands. In the three samples studied, ice appears as a continuous network rather than as single crystals. This network structure is better visible on the single-phase than on the multi-phase pictures, due to the higher depth of field. Small ice blocks with a diameter of 1-10 µm, integrated in the ice network, are observed in some of the spaces between the starch granules. Yeast cells appear on the picture as ellipsoids with a size of 4-5 µm, homogeneously distributed within the samples. The yeast images also show a background noise, especially in the gluten-rich regions. Liquid water appears to be present in the areas where no other phase is present.

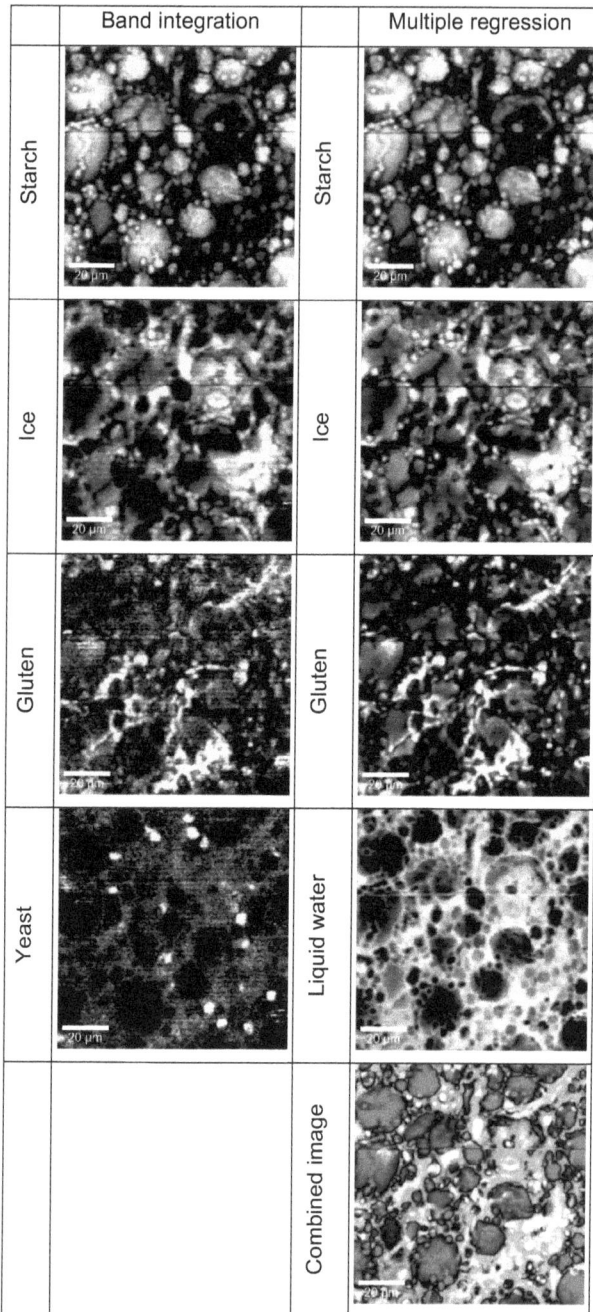

	Band integration		Multiple regression
Starch		Starch	
Ice		Ice	
Gluten		Gluten	
Yeast		Liquid water	
		Combined image	

Figure 2.2: First sample: Distribution of the dough components according to both data processing methods (left: band integration; right: multiple regression). Colour code of the bottom image: starch = red, gluten = yellow, ice = white, liquid water = green.

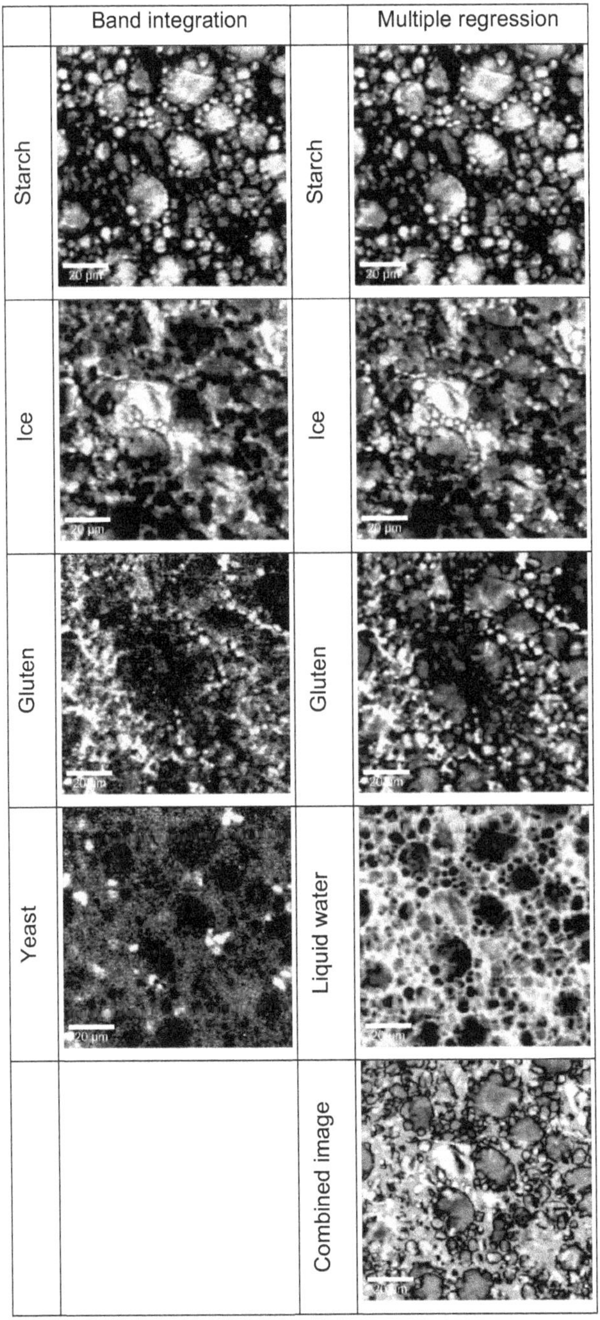

Figure 2.3: Second sample: Distribution of the dough components according to both data processing methods. Colour code of the bottom image: starch = red, gluten = yellow, ice = white, liquid water = green.

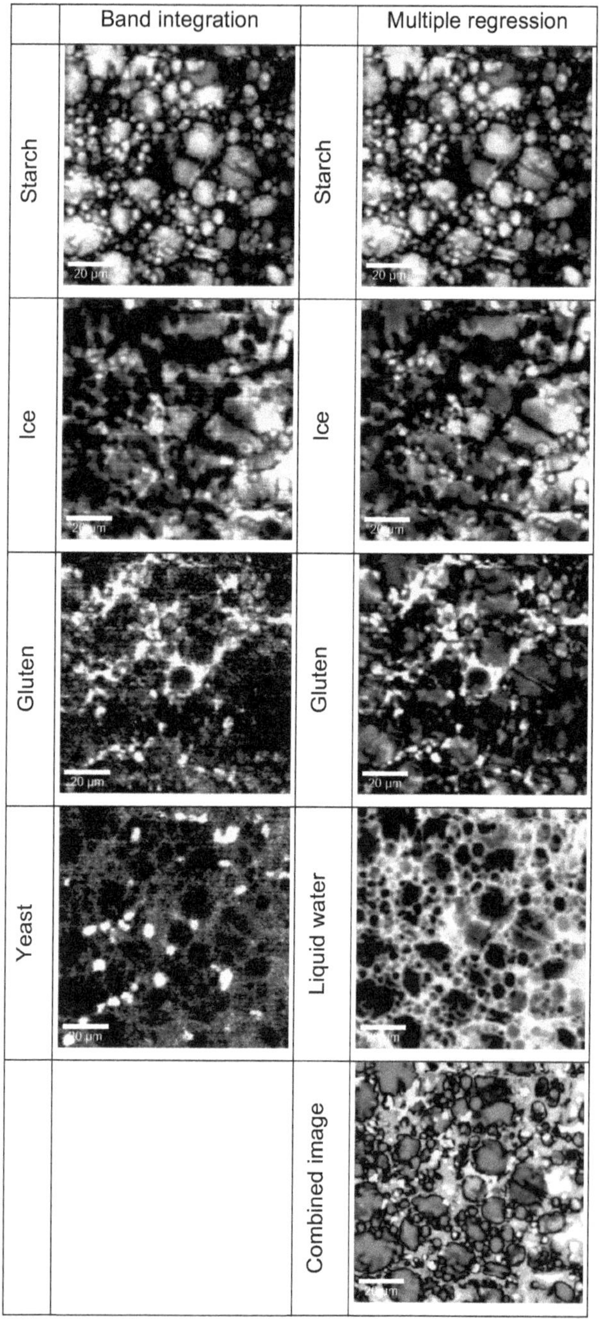

Figure 2.4: Third sample: Distribution of the dough components according to both data processing methods. Colour code of the bottom image: starch = red, gluten = yellow, ice = white, liquid water = green.

Discussion

Sample integrity

In Raman microscopy, in order to obtain a detectable signal, a laser beam with high power density needs to be applied in the focus area, which can result in heating and structural alteration. Especially with frozen samples it is therefore essential to ascertain the integrity of the samples after the measurements. In the case of the samples of the present study, routine microscopic inspection of the Raman-mapped sample areas showed no signs of damage. Although it is not possible to measure temperature within the sample during measurement, two observations suggest that temperature was not significantly increased: (1) Ice and liquid water were found to be both strongly represented in the investigated samples; this is consistent with the DSC measurements of frozen dough by Baier-Schenk et al. (2005a), which show that at -15 C about 50% of the water is in the frozen state, whereas the other 50% is in the liquid form; (2) Repeated mappings of the same areas yielded identical results; under the assumption of melting and recrystallization, a different distribution of ice would have been observed.

Band position and assignment

The spectral bands used for imaging in the first method were chosen both by (1) comparing the spectra of the single components on Fig. 1 and searching for bands that are unique for each component and (2) using knowledge from literature on band assignment. In the case of the OH stretching band of ice and the stretching vibration of the carbon network of starch, the high intensity of the bands and their characteristic shape allows for a clear assignment. These bands are very appropriate for identifying ice and starch in the frozen dough. The assignment of the amide I band is more complicated due to the fact that the shape of the band and the position of its maximum depend on the secondary and tertiary structure of the proteins (Tuma, 2005). A further difficulty lies in the proximity of other bands. Finally, fluorescence, which is dependent on the excitation wavelength, overlaps with the Raman signal. For these reasons, there is no certitude that the spectral range selected (16451690 cm^{-1}) corresponds exclusively to amide I in gluten. In addition, it must be noted that amide I can only be seen as an imperfect indicator of gluten in the frozen dough system, as amide I signal is also expected to arise from non-gluten wheat protein and from yeast protein.

41

The use of the band 740-766 cm^{-1} for yeast identification in the dough must be considered as an empirical approach. It is unclear whether the signal measured in this range is solely attributable to the ring breathing vibration of tryptophan, nor whether tryptophan can be considered as a reliable indicator of yeast in the dough system.

Unambiguous identification and imaging of the single dough components

As discussed above, starch and ice have good single band indicators in the frozen dough system, and it is not surprising that for these components both data processing methods lead to similar pictures. The pictures generated by multiple linear regression show less noise and are therefore sharper, probably due to the fact that they are based on a broader data basis. In the case of gluten, a good match between the pictures obtained from both data processing methods is observed as well.

The identification and imaging of yeast has a lower level of confidence, due to the limitations described above. The noise observed on the images can be explained by the fact that gluten and starch also have weak bands in the chosen spectral range. Imaging of yeast using multiple linear regression was not successful, as the generated images were obviously wrong (no cell shape); this is probably due to the overlap of the yeast signals with signals from the other dough components in most spectral areas, as well as to the low abundance of yeast in the system.

Liquid water can be identified only by multiple linear regression, due to the overlap with the OH bands of the other components.

Spatial distribution of the single dough components

The observed spatial distribution of starch, gluten and yeast is consistent with data from literature. A bimodal size distribution of starch granules in wheat flour was reported by numerous authors like Stoddard (1999). The observed relative distribution of starch and gluten, with gluten fibrils organised as a network around the starch granules, is consistent with observations made by other techniques like scanning electron microscopy (Yi and Kerr, 2009), confocal scanning laser microscopy and epifluorescence light microscopy (Peighambardoust et al., 2010). The observed size and shape of the yeast cells is in line with literature data (Smith et al., 2000).

The most interesting, and really novel aspect, is the distribution of ice. The structure of ice as a continuous phase (crystal network) within the frozen dough has not, to our knowledge, been reported elsewhere so far. This continuous structure may be of importance for understanding damage to the other dough components, especially to gluten which also has a network structure – meaning that in the frozen state, the gluten network and the ice crystal network coexist and are embedded in one another.

Conclusions

In our investigations, confocal Raman microscopy allowed a reliable identification and imaging of starch, ice and gluten; yeast and liquid water were identified with a lower degree of confidence. The method is non-destructive and does not require any staining.

The unambiguous identification of ice based on its specific Raman spectrum (specific OH stretching band) allows visualising the structure of ice within the frozen dough matrix. The structure of ice as a network rather than isolated crystals represents a new finding that helps understanding the interactions between the dough components in the frozen state.

We suggest that the technique described in this paper may be useful to study the influence of different freezing and storage conditions, of different storage times, and of specific ingredients such as ice structuring proteins, on the ice network structure in frozen dough. Such investigations may be conducted either on a model system like in this study (dough frozen on a microscope slide), or on microtome sections of real-life frozen products.

The technique in itself may be refined in terms of spatial resolution by the use of an objective with a higher magnification and in terms of measurement speed by the use of a more sensitive spectrometer. The use of a different excitation wavelength could help reducing fluorescence. More detailed Raman spectroscopic studies of the single components of the dough, especially starch and gluten, may allow differentiating between sub-components such as amylose, amylopectin, gliadin and glutenin, ultimately leading to more detailed images.

Numerous further applications of cryo Raman microscopy are conceivable with other kinds of frozen foods or frozen biological samples.

References

Baier-Schenk, A., Handschin, S., Conde-Petit, B., 2005a. Ice in prefermented frozen bread dough – an investigation based on calorimetry and microscopy. Cereal Chem. 82, 251-255.

Baier-Schenk, A., Handschin, S., von Schönau, M., Bittermann, A.G., Bächi, T., Conde-Petit, B., 2005b. In situ observation of the freezing process in wheat dough by confocal laser scanning microscopy (CLSM): formation of ice and changes in the gluten network. J. Cereal Sci. 42, 255-260.

Berglund, P.T., Shelton, D.R., Freeman, T.P., 1991. Frozen bread dough ultrastructure as affected by duration of frozen storage and freeze-thaw cycles. Cereal Chem. 68, 105-107.

Dieing, T., Hollricher, O., Toporski, J., 2011. Confocal Raman Microscopy. Springer, Heidelberg.

Đuričković, I., Claverie, R., Bourson, P., Marchetti, M., Chassot, J.M., Fontana, M.D., 2011. Water-ice phase transition probed by Raman spectroscopy. J. Raman Spectrosc. 42, 1408-1412.

Esselink, E.F.J., van Aalst, H., Maliepaard, M., van Duynhoven, J.P.M., 2003. Longterm storage effect in frozen dough by spectroscopy and microscopy. Cereal Chem. 80, 396-403.

Fechner, P.M., Wartewig, S., Kiesow, A., Heilmann, A., Kleinebudde, P., Neubert, R.H.H., 2005. Influence of water on molecular and morphological structure of various starches and starch derivatives. Starch/Stärke 57, 605-615.

Jääskeläinen, A.S., Holopainen-Mantila, U., Tamminen, T., Vuorinen, T., 2013. Endosperm and aleurone cell structure in barley and wheat as studied by optical and Raman microscopy. J. Cereal Sci. 57, 543-550.

Krafft, C., Dietzek, B., Schmitt, M., Popp, J., 2012. Raman and coherent anti-Stokes Raman scattering microspectroscopy for biomedical applications. J. Biomed. Opt. 17, 040801.

Le Bail, A., Zia, C., Giannou, V., 2012. Quality and safety of frozen bakery products. In: Sun, D.W. (Ed.), Handbook of Frozen Food Processing and Packaging, second ed. CRC Press, Boca Raton, pp. 501-528.

Peighambardoust, S.H., Dadpour, M.R., Dokouhaki, M., 2010. Application of epifluorescence light microscopy (EFLM) to study the microstructure of wheat dough: a comparison with confocal scanning laser microscopy (CSLM) technique. J. Cereal Sci. 51, 21-27.

Piot, O., Autran, J.C., Manfait, M., 2000. Spatial distribution of protein and phenolic constituents in wheat grain as probed by confocal microspectroscopy. J. Cereal Sci. 32, 57-71.

Piot, O., Autran, J.C., Manfait, M., 2001. Investigation by confocal Raman microspectroscopy of the molecular factors responsible for grain cohesion in the Triticum aestivum bread wheat. Role of the cell walls in the starchy endosperm. J. Cereal Sci. 34, 191-205.

Piot, O., Autran, J.C., Manfait, M., 2002. Assessment of cereal quality by micro-Raman analysis of the grain molecular composition. Appl. Spectrosc. 56, 1132-1138.

Rösch, P., Harz, M., Peschke, K.-D., Ronneberg, O., Burkhardt, H., Popp, J., 2006. Identification of single eukaryotic cells with micro-Raman spectroscopy. Biopolymers 82, 312-316.

Smith, E., Dent, G., 2005. Modern Raman Spectroscopy: a Practical Approach.Wiley, Chichester.

Smith, A.E., Zhang, Z., Thomas, C.R., Moxham, K.E., Middelberg, A.P., 2000. The mechanical properties of Saccharomyces cerevisiae. Proc. Natl. Acad. Sci. 97, 9871-9874.

Stoddard, F.L., 1999. Survey of starch particle-size distribution in wheat and related species. Cereal Chem. 76, 145-149.

Tuma, R., 2005. Raman spectroscopy of proteins: from peptides to large assemblies. J. Raman Spectrosc. 36, 307-319.

Weikusat, C., Freitag, J., Kipfstuhl, S., 2012. Raman spectroscopy of gaseous inclusions in EDML ice core: first results e microbubbles. J. Glaciol. 58, 761-766.

Yi, J., Kerr, W.L., 2009. Combined effects of freezing rate, storage temperature and time on bread dough and baking properties. LWT – Food Sci. Technol. 42, 1474-1483.

Zounis, S., Quail, K.J., Wootton, M., Dickson, M.R., 2002. Effect of final dough temperature on the microstructure of frozen bread dough. J. Cereal Sci. 36, 135-146.

Chapter 3 – Wheat flour quality evaluation from the baker's perspective: comparative assessment of 18 analytical methods

Abstract

In this study, we sourced 37 commercial flours from 14 mills based on 7 countries and analysed them with a total of 18 methods, generating 90 single analytical values for each flour. The 18 methods were chosen to cover the analytical practice of most European mills and bakery companies, as well as particle charge detection, GlutoPeak and solvent retention capacity as emerging methods. We investigated the relationship between the data from the individual methods, and performed a principal component analysis to describe the structure of the data set and identify the main underlying flour properties. Four principal components accounted for 64.8% of the total variance. They were interpreted as (PC1) starch gelatinization properties, (PC2) hydration properties, (PC3) dough resistance at variable water amount, and (PC4) dough strength at fixed water amount. From the emerging methods, solvent retention capacity (sodium carbonate and water) was highly correlated with PC2, while the GlutoPeak max torque was highly correlated with PC4.

Keywords

Wheat flour, Principal component analysis, Rheology, Functional properties

---

This chapter has been published:

Huen J, Börsmann J, Matullat I, Böhm L, Stukenborg F, Heitmann M, Zannini E, Arendt EK (2018). Wheat flour quality evaluation from the baker's perspective: comparative assessment of 18 analytical methods. *European Food Research and Technology*, 244 (3), 535-545. doi:10.1007/s00217-017-2974-3

Introduction

Both industrial and craft bakeries are facing the problem of variability of wheat flour quality. This variability makes it impossible to bake over a longer period of time with constant recipes and constant process parameters. To be able to react pro-actively to flour quality variations, bakery companies need reliable analytical methods that allow predicting the behaviour of flour in production as well as the final bakery results. In practice, flour analysis is usually performed in the mill; bakery companies rely on values provided on analytical certificates, which typically include 5–10 parameters.

In this context, millers and bakers need to agree on a set of measurements that is practicable in daily production and delivers the most useful information. Interestingly, across Europe, the analytical practice differs depending on the countries. In UK and Ireland, for example, the level of damaged starch is considered to be of high importance for functionality. In France and Spain, AlveoLAB measurements are considered as a standard. In Germany, Farinogram measurements are common. One could argue that the reason for this different practice lies in different applications (local bread recipes), but a part of the explanation certainly lies in historical reasons, habits, and the continuous use of equipment already available in the mills.

Globalization of wheat commerce, modification of wheat quality related to climate change [520], evolution in bakery production practice (industrialisation, increasing use of freezing technology [3, 11]) , change in consumption habits, and the availability of novel flour analysis methods are good reasons for questioning the flour analytical practice.

In this study, our goal was to compare and better understand the information provided by different flour analysis methods used across Europe, including both established and emerging methods. To this end, we analysed 37 commercial flours from 14 mills located in 7 European countries by 18 different methods, generating 90 analytical values for each flour. The similarities and differences of the data obtained by the different methods were evaluated by statistical techniques.

Table 3.1 gives an overview of the methods used in our study. Some methods quantify the amount of specific flour components (water, minerals, protein, acids, and damaged starch), while other characterize the functionality of gluten, starch, pentosans, and alpha-amylases. The table gives an evaluation of the level of skills required, the price category of the equipment, the workload per sample, and the analysis time.

Table 3.1: Comparison of the methods used in the study.

Method / Device	Standard	Measurement of substance concentration	Functionality assessment of...	Price category of equipment	Level of skills required	Approximate workload per sample (min)	Approximate analysis time (min)
Moisture	ICC 110/1	Water		•	•	15	180
Ash content	ICC 104/1	Minerals		•	••	40	300
Protein content	ICC 105/2	Protein		••	•••	60	240
Sedimentation value	ICC 116/1		Gluten	•	••	15	45
Brabender GlutoPeak	setting: 25°C, 3000 rpm		Gluten	•••	•	10	15
Perten Glutomatic	ICC 155	Wet gluten Dry gluten	Gluten	••	•••	30	30
pH and acidity	ICC 145	H_3O^+, Acids		•	••	30	40
Falling number	ICC 107/1		Endogenous alpha-amylases	••	•	10	20
Perten Rapid Visco Analyser	-		Starch (Gluten)	••••	•	10	15
Brabender Micro Visco Amylograph	-		Starch (Gluten)	•••	••	15	00
Chopin SDmatic	ICC 172	Damaged starch		••	•	15	15
SRC-CHOPIN	AACC 56-11		Damaged starch, Glutenin, Pentosans	••••	•	15	30
Mütek Particle Charge Analyser	- Schick (2010)		Particle charge of macroions	••	••	75	75
Brabender Farinograph	ICC 115/1		Gluten, Starch	••••	•	15	45
Chopin Mixolab	ICC 173		Gluten, Starch	••••	•	30	60
Brabender Extensograph	ICC 114/1		Gluten	•••	•	60	180
Chopin AlveoLAB	ICC 121		Gluten	••••	•	20	50
Chopin Rheo F4	AACC 89-01.01		Gluten	••	•	20	200

Experimental

Raw materials

37 wheat flours (50 kg of each) were sourced from 14 European mills located in Belgium, France, Germany, Italy, The Netherlands, Spain, and the United Kingdom. The flours were specified regarding their ash content (0.50–0.65 after ICC 104), their protein content (min 11.0 after ICC 167), and their moisture content (max. 16.0 after ICC 110/1). In some countries, it is usual to add some substances to the flour in the mill during production (e.g., gluten, enzymes, malt flour, vitamins, and minerals). Our flours were packed without these additions at our demand. The only additive we accepted was ascorbic acid, which is used widely across Europe and was contained at concentrations between 0 and 25 ppm in the assessed flours.

Analytics

All 37 flours were analysed by a series of methods that were selected to reflect the diversity of analytical approaches commonly used by millers and bakers across Europe. The corresponding ICC and AACC standards are mentioned in Tables 3.1 and 3.2. The selected methods included moisture, ash, protein, wet and dry gluten, and damaged starch content as well as pH and acidity. Furthermore, functionality assessment was performed with the established methods sedimentation value, falling number, Brabender (Duisburg, Germany) Farinograph, Extensograph and Micro Visco Amylograph, Chopin (Villeneuve-la-Garenne, France) Mixolab, AlveoLAB and Rheo F4, as well as Perten (Hägersten, Sweden) Glutomatic and Rapid Visco Analyser. Finally, the following emerging methods and devices were used: Brabender GlutoPeak, Chopin SRC-CHOPIN, and BTG (Eclépens, Switzerland) Mütek PCD-05. GlutoPeak is a gluten aggregation test, giving an indication of the gluten strength (aggregation energy GPG) and of the speed of aggregation (time of maximum torque GPT). Several authors reported correlations of the GlutoPeak values with Alveograph, Extensograph, and Farinograph values ([16], [24]) as and with the level of protein and gluten [17]. Compared to those methods, the main advantages of GlutoPeak are the speed of analysis and the lower price of the device. With the SRC-CHOPIN based on the solvent retention capacity method [13], the functionality of gluten, pentosans and damaged starch, as well as flour hydration can be investigated in one device. Particle charge detection investigates the electrical load of macroions in a flour suspension and represents a different, innovative approach to flour functionality [19].

A code was assigned to each parameter, as shown in Table 3.2. For each flour, all analytical investigations were performed within 2 weeks, to exclude as far as possible the influence of quality change over time (as described in [8]) on the correlations observed. The acid ascorbic content ASA as declared by the mills was also included into the data set.

Table 3.2: Methods used for flour analytics in WP2 and results obtained with the 37 248 investigated flours (mean, min, max, relative standard deviation). R = number or repetitions. dl = dimensionless, mu =250 device manufacturer unit

Method / Device	Standard	R	Output	Code	Mean	Min	Max	rsd	Unit
Moisture	ICC 110/1	3	Moisture after 90 min	MOI	0,13	0,11	0,15	6,1%	% (dl)
Ash content	ICC 104/1	2	Ash in dry substance	ASH	0,57	0,44	0,70	10,0%	% (dl)
Protein content	ICC 105/2	2	Protein in dry substance	PRT	12,2	10,1	13,9	6,9%	% (dl)
Sedimentation value	ICC 116/1	2	Sedimentation value	SDV	34,3	24,8	53,4	16,2%	mL
Brabender GlutoPeak	setting: 25°C, 3000 rpm	2	Time of max torque	GPT	72,7	46,5	124,0	28,1%	s
			Max Torque	GPM	81,2	67,0	91,5	8,9%	BU (mu)
			Torque before maximum	GPB	34,3	23,5	44,0	16,4%	BU (mu)
			Torque after maximum	GPA	60,6	46,5	75,5	9,3%	BU (mu)
			Startup energy	GPS	277	148	366	18,0%	GPE (mu)
			Aggregation energy	GPG	1916	1512	2329	9,7%	GPE (mu)
Perten Glutomatic	ICC 155	2	Wet gluten	GMW	27,7	21,8	33,0	10,3%	% (dl)
			Dry gluten	GMD	10,6	7,9	13,3	10,5%	% (dl)
			Gluten index	GMI	84,4	62,6	94,4	9,9%	% (dl)
pH and acidity	ICC 145	2	pH	PH	6,3	6,0	7,6	4,7%	- (dl)
			Acidity	ADT	0,97	0,80	1,21	10,6%	acidity unit
Falling number	ICC 107/1	5	Falling number	FAN	356	277	419	11,4%	s
Perten Rapid Visco Analyser	-	3	Peak viscosity	VAM	2227	1498	2910	15,0%	RVU (mu)
			Holding viscosity	VAY	1299	611	1776	25,0%	RVU (mu)
			Breakdown viscosity	VAB	928	761	1186	10,5%	RVU (mu)
			Final viscosity	VAV	2348	1303	3061	19,6%	RVU (mu)
			Setback viscosity	VAS	1050	691	1337	13,9%	RVU (mu)
			Peak time	VAT	6,11	5,56	6,38	3,2%	s
			Pasting temperature	VAP	59,9	58,1	62,5	1,9%	°C
			Peak temperature	VAX	95,1	95,0	95,2	0,1%	°C
Brabender Micro Visco Amylograph	-	2	Hot viscosity	MVB	633	316	1138	32,5%	mPas
			Cold viscosity	MVE	1491	922	2531	24,8%	mPas
			Temperature of start of gelatinisation	MTA	59,2	54,7	62,7	3,2%	°C
			Temperature of max viscosity	MTB	89,1	83,3	91,7	2,6%	°C
			Viscosity at start of cooling	MVD	417	139	819	39,7%	mPas
			Viscosity at start of holding	MVC	595	221	1123	38,3%	mPas
			Difference Viscosity B-D	MVX	216	137	366	24,4%	mPas
			Difference Viscosity E-D	MVY	1074	773	1712	20,7%	mPas
Chopin SDmatic	ICC 172	1	Damaged starch (AI)	DST	94,4	93,2	95,9	0,8%	%AI (dl)
SRC-CHOPIN	AACC 56-11	1	SRC Water	SRW	65,6	55,4	73,8	6,9%	% (dl)
			SRC Sucrose	SRS	105	97	116	4,7%	% (dl)
			SRC Lactic Acid	SRL	124	97	156	8,9%	% (dl)
			SRC Sodium Carbonate	SRC	83,1	68,2	94,1	7,3%	% (dl)
Mütek Particle Charge Analyser	Schick (2010)	2	Total charge after 4 min swelling	P04	-1,52	-2,70	-0,24	-35,5%	C.g^{-1}
			Total charge after 24 min swelling	P24	-1,46	-2,66	-0,19	-36,1%	C.g^{-1}
			Total charge after 44 min swelling	P44	-1,22	-2,51	0,05	-47,4%	C.g^{-1}
			Total charge after 64 min swelling	P64	-0,84	-2,32	0,13	-70,7%	C.g^{-1}
Brabender Farinograph	ICC 115/1	2	Water absorption 500 BE	FWA	58,1	50,9	61,7	4,0%	% (dl)
			Dough development time	FDD	2,03	1,36	4,63	28,0%	min
			Stability	FST	9,80	2,83	22,17	42,9%	min
			Dough softening	FDS	61,1	21,0	107,0	40,0%	BE
			Farino quality value	FQV	82,4	29,0	181,5	47,2%	- (dl)

			Hydration	XHY	59,4	52,4	62,3	3,4%	% (dl)
Chopin Mixolab	ICC 173	1	C1 Time	X1T	3,36	1,08	8,17	69,0%	min
			C1 Torque	X1D	1,10	1,05	1,14	2,4%	Nm
			C1 Dough temperature	X1C	30,0	28,2	31,5	2,5%	°C
			C2 Time	X2T	16,7	16,0	17,6	2,1%	min
			C2 Torque	X2D	0,51	0,34	0,59	9,4%	Nm
			C2 Dough temperature	X2C	52,8	50,7	55,5	2,1%	°C
			C3 Time	X3T	23,5	22,7	29,1	4,7%	min
			C3 Torque	X3D	2,04	1,70	2,31	5,8%	Nm
			C3 Dough temperature	X3C	78,0	75,3	84,8	2,1%	°C
			C4 Time	X4T	31,7	27,3	35,2	5,2%	min
			C4 Torque	X4D	1,78	1,29	2,26	13,0%	Nm
			C4 Dough temperature	X4C	85,1	80,5	87,1	2,1%	°C
			C5 Time	X5T	45,0	45,0	45,0	0,0%	min
			C5 Torque	X5D	2,68	1,81	3,74	16,0%	Nm
			C5 Dough temperature	X5C	58,9	60,2	57,2	1,4%	°C
			Amplitude	XAM	0,09	0,01	0,11	22,2%	Nm
			Stability	XST	9,71	5,83	11,27	11,2%	min
Brabender Extensograph	ICC 114/1	2	Energy 45 min	EE45	108	74	165	19,7%	cm²
			Resistance 45 min	ER45	289	199	405	17,0%	BE (mu)
			Extensibility 45 min	EX45	189	166	310	13,9%	Mm
			Maximum 45 min	EM45	406	225	581	20,5%	BE (mu)
			Ratio 45 min	EY45	1,59	0,99	2,75	23,4%	- (dl)
			Db/Max 45 min	ED45	2,25	1,41	3,74	22,9%	- (dl)
			Energy 90 min	EE90	113	59	193	23,6%	BE (mu)
			Resistance 90 min	ER90	343	202	502	24,5%	Mm
			Extensibility 90 min	EX90	171	78	216	13,4%	BE (mu)
			Maximum 90 min	EM00	400	247	720	20,2%	- (JI)
			Ratio 90 min	EY90	2,03	1,17	3,40	32,0%	- (dl)
			Db/Max 90 min	ED90	2,87	1,22	4,65	32,5%	BE (mu)
			Energy 135 min	EE135	115	66	176	23,6%	Mm
			Resistance 135 min	ER135	355	212	559	26,3%	BE (mu)
			Extensibility 135 min	EX135	172	141	216	9,5%	- (dl)
			Maximum 135 min	EM135	490	259	769	27,9%	- (dl)
			Ratio 135 min	EY135	2,12	1,16	3,83	33,9%	- (dl)
			Db/Max 135 min	ED135	2,90	1,52	4,89	33,5%	- (dl)
Chopin AlveoLAB	ICC 121	1	Tenacity P	AVP	93,7	36,0	119,0	19,9%	mmH₂O
			Extensibility L	AVL	91,0	61,0	157,0	24,7%	m
			Extensibility G	AVG	21,1	17,4	27,9	11,8%	- (dl)
			Baking strength W	AVW	274	141	419	20,7%	10⁻⁴ J
			Ratio P/L	AVR	1,12	0,23	1,95	34,6%	- (dl)
			Elasticity Ie	AVI	55,3	47,7	65,8	8,0%	% (dl)
Chopin Rheo F4	AACC 89-01.01	1	Dough development Hm	RDM	53,0	44,9	75,4	10,4%	mm
			Dough development h	RDH	51,0	41,8	75,4	11,3%	mm
			Dough development (Hm-h)/Hm	RDR	3,68	0,00	17,20	132,7%	% (dl)
			Dough development T1	RD1	0,11	0,06	0,13	16,3%	H
			Gaseous release H'm	RGH	72,2	60,0	76,8	4,3%	Mm
			Gaseous release T'1	RG1	0,05	0,04	0,08	17,3%	H
			Gaseous release Tx	RGX	0,05	0,04	0,06	13,3%	H
			Gaseous release Vt	RGT	1625	1383	1767	4,9%	mL
			Gaseous release Vr	RGR	1270	1196	1347	3,1%	mL
			Gaseous release Vc	RGC	355	151	461	19,0%	mL
			Gaseous release Vr/Vt	RGS	78,3	72,9	89,7	4,4%	% (dl)
Ascorbic Acid	Spec	-	Ascorbic acid content	ASA	7,27	0	25	121%	ppm

Data analysis

The Pearson's correlation coefficients relating the individual parameters were calculated, with the corresponding P values. The structure of the data set was further investigated by carrying out a principal component analysis with a Varimax rotation on four principal components. All calculations were performed using IBM SPSS Statistics version 22.

Results

Overview of data

Table 3.2 gives an overview of the data obtained, with the minimum and maximum values observed for each parameter, the average, and the relative standard deviation. Some values were found to be slightly out of our specification, but we chose to keep the corresponding samples in the evaluation, as they represent the quality available on the market.

Observed correlations

Table 3.3 gives all correlation coefficients with an absolute value superior or equal to 0.7 and a P value below 0.001, which were calculated between parameters generated by different methods. Table 3.4 gives the results of the PCA, whose four first principal components account for 64.8% of the variance in the data set. Figures 3.1, 3.2 show loadings plots of the single analytical parameters on the four principal components. In the following, the main relationships observed are described following the structure given by the PCA.

Table 3.3: Correlation coefficients with absolute values higher than 0,7 and P < 0,001, between parameters from different methods.

(A)

		Rapid Visco Analyser				
		VAM	VAY	VAV	VAS	VAT
Falling number	FAN	,735	,793	,809	,781	,794
Mixolab	X4T	-,710	-,812	-,819	-,770	-,732
	X4D	,803	,794	,807	,770	
	X5D	,849	,854	,892	,905	

(B)

		SRC-CHOPIN			Water Absorption	
		SRW	SRC	SRS	FWA	XHY
Particle Charge Analyser	P04	-,739	-,775			
	P24	-,703	-,736			
Damaged Starch	DST	,792	,866	,797		
Water absorption	XHY	,742	,757		,869	
AlveoLAB	AVP	,753	,738		,823	,945
	AVL		-,792			-,752
	AVG		-,786			-,735
	AVR	,716	,746			,804
RVA	VAP	-,721	-,726			

(C)

		Extensograph		
		ED90	EY135	ED135
Ascorbic acid	ASA	,709	,714	,719

(D)

		AlveoLAB			SRC-Chop.	Water absorption		Protein
		AVP	AVW	AVI	SRL	FWA	XHY	PRT
Protein	PRT		,736	,777				
Sedi. value	SDV				,819			
GlutoPeak	GPM		,745					,787
	GPB					,796		
	GPA	,708	,707			,777	,721	
	GPS	,731				,701	,705	
	GPG	,706				,786	,714	
Extensograph	EE45		,767	,858				
	EE90		,749	,804				
	EM90			,712				
	EE135		,828	,877				,706
	EM135			,729				
Glutomatic	GMD							,738

(E)

		Mixolab			AlveoLAB
		X1T	X2D	XST	AVW
Farinograph	FST		,796	,812	,703
	FDS		-,847	-,836	
	FQV	,733	,707	,798	

Table 3.4: Loading values of the single analytical parameters on the first 4 principal components after Varimax rotation (values > 0,7).

	PC1		PC2		PC3		PC4
VAV	,937	SRC	-,882	ED90	,951	AVW	,888
VAY	,930	DST	-,851	ED135	,942	AVI	,783
MVD	,909	SRW	-,838	ER135	,932	PRT	,756
MVC	,905	AVL	,821	EY135	,920	EX45	,749
VAS	,878	AVG	,820	ED45	,914	GPM	,702
MVB	,865	AVR	-,798	EY90	,910		
X5D	,865	XHY	-,760	EM135	,895		
VAM	,862	AVP	-,754	ER90	,893		
VAT	,838	RGC	-,745	ER45	,884		
FAN	,825	MTA	,741	EM90	,863		
X4T	-,800	RGS	,739	EY45	,830		
MVE	,784	P04	,729	EM45	,808		
X4D	,777	VAP	,710	ASA	,778		
MTB	,775	SRS	-,708				
X4C	,761						
XST	,740						

The first principal component is related to rapid visco analyser values (peak viscosity VAM, holding viscosity VAY, setback viscosity VAS, and final viscosity VAV), micro visco amylograph values (viscosity at start of cooling MVD, viscosity at start of holding MVC, hot viscosity MVB, and cold viscosity MVE), Mixolab values (C4 torque X4D, C5 torque X5D, C4 time X4T, and stability XST), and the falling number FAN. Referring to Table 3.3A, the rapid viso analyser values have strong and significant correlations both with the falling number FAN and the Mixolab C4 and C5 torque X4D, X5D and C4 time X4T.

The second principal component is related to solvent retention capacity values (calcium carbonate SRC, water SRW), the level of damaged starch DST, some AlveoLAB values (extensibility AVL, AVG, ratio AVR, and tenacity AVP), Mixolab hydration XHY, Rheo F4 gaseous release RGC and RGS, particle charge at 4 min of swelling P04, and the temperature at start of gelatinization from the micro visco amylograph MTA and the rapid visco analyser VAP. Referring to Table 3.3B, the SRC water value SRW correlates with some PCD values (P04, P24), the damaged starch level DST, the Mixolab hydration value XHY, the AlveoLAB tenacity AVP and ratio AVR, and the pasting temperature VAP as measured by the rapid visco analyser. The solvent retention capacity calcium carbonate value SRC correlates with all the previous ones and the AlveoLAB AVL and AVG values. Next to this, the Mixolab hydration value

XHY correlates with the Farinograph water absorption FWA as well as with all AlveoLAB values. The Farinograph water absorption FWA also correlates with the dough tenacity AVP as measured by the AlveoLAB.

The third principal component is almost solely related to some Extensograph values (resistance ER45, ER90, and ER135, resistance maximum EM45, EM90, and EM135, ratio resistance/extensibility EY45, EY90, and EY135, and ratio resistance maximum/extensibility ED45, ED90, and ED135), only with the amount of ascorbic acid ASA also having a high value on it (above the threshold of 0,7). Referring to Table 3.3C, the amount of ascorbic acid ASA is related to the Extensograph ratio values ED90, ED135, and EY135.

The fourth principal component is related to further AlveoLAB values (baking strength AVW and index AVI), the protein content PRT, the Extensograph extensibility at 45 min EX45 and the GlutoPeak max torque GPM. Referring to Table 3.3D, the AlveoLAB baking strength AVW and index AVI correlate with the protein content PRT. The baking strength AVW also correlates with the GlutoPeak max torque GPM.

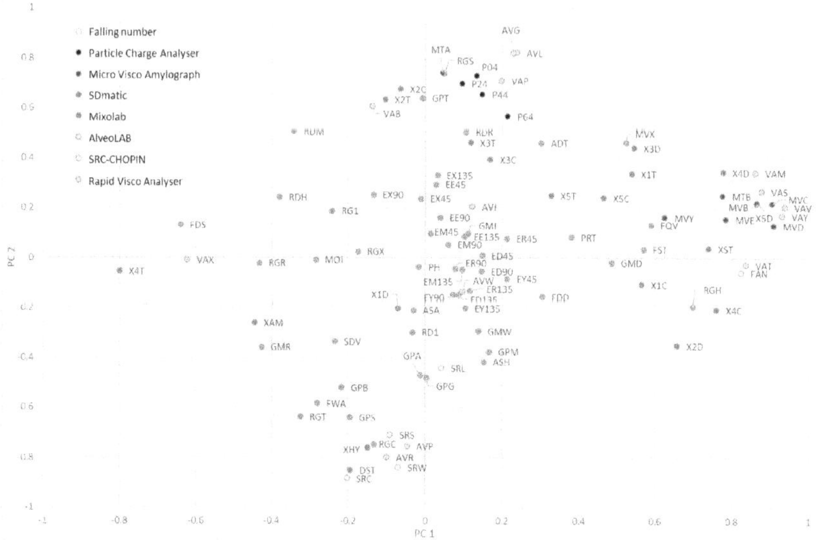

Figure 3.1: Loadings plot of the single analytical parameters on PC1 and PC2. The methods that reached values higher than 0.7 on PC1 or PC2 are shown in a specific color and mentioned in the legend.

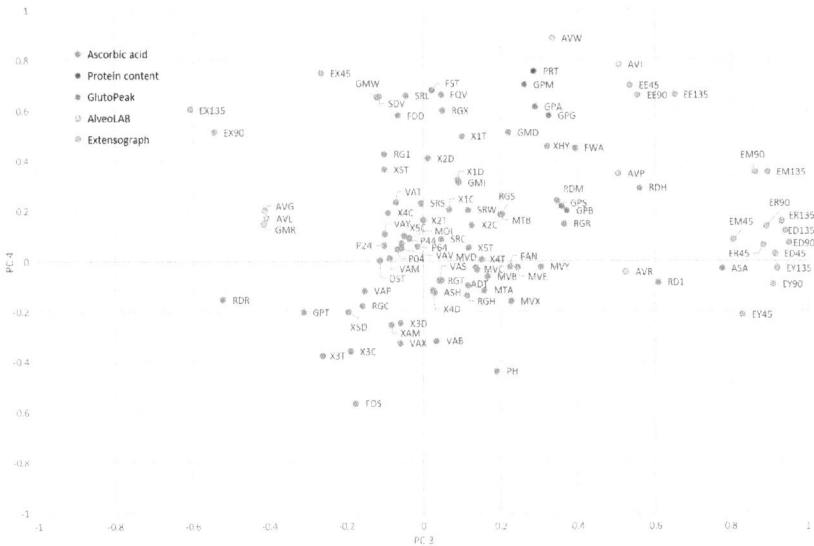

Figure 3.2: Loadings plot of the single analytical parameters on PC3 and PC4. The methods that reached values higher than 0.7 on PC4 or PC4 are shown in a specific color and mentioned in the legend.

In addition to the structure given by the principal component analysis, Table 3.3D presents further interesting correlations. Indeed, the AlveoLAB baking strength AVW is also related to the Extensograph energy values EE45, EE90, and EE135, the GlutoPeak torque at maximum GPM, and torque after maximum GPA. The AlveoLAB index value AVI also correlates with the Extensograph energy and maximum values EE45, EE90, EE135, EM90, and EM135. Furthermore, the AlveoLAB tenacity AVP is related to the GlutoPeak start energy GPA, aggregation energy GPS, and torque after maximum GPG. Next to this, the SRC lactic acid value SRL correlates with the sedimentation value SDV. The Farinograph and Mixolab hydration values FWA and XHY correlate with the GlutoPeak start energy GPS, aggregation energy GPG, and torque after maximum GPA. Finally, the dry gluten GMD as determined with the Glutomatic correlates with the protein content PRT.

Other relationships are described in Table 3.3E: The Farinograph stability FST correlates with the AlveoLAB baking strength AVW, the Mixolab stability XST, and the

Mixolab C2 torque X2D. The Farinograph dough-softening FDS correlates negatively with the Mixolab stability XST and the Mixolab C2 torque X2D. The Farinograph quality value FQV is related to the Mixolab stability XST, the Mixolab C2 torque X2D, and the Mixolab C1 time X1T.

Parameters without significant correlations

It is remarkable that the ash content ASH, the pH value PH, the acidity ADT, most of the Rheo F4 values, and the Gluten Index GMI and Wet Gluten GMW level were found to have no high and significant correlations with the other parameters measured. This indicates that these parameters are, to a certain extent, independent from the other flour properties.

Discussion

Starch gelatinisation and viscosity

PC1 is related to starch gelatinisation and viscosity. The methods rapid visco analyser, micro visco amylograph, Mixolab, and falling number are all based on forming a starch gel by following a temperature cycle in the presence of water. The measured viscosities are mainly related to starch gelatinization, starch degradation by endogenous alpha-amylases and starch retrogradation. Similar correlations were reported by Deffenbaugh and Walker [6] (RVA vs. micro visco amylograph) as well as Peña and Posadas-Romano [18] (Mixolab C4 and C5 vs. Falling number).

No relation with analytical values describing gluten was observed on PC1 (at the threshold of 0,7), suggesting that the viscosity values represented on this principal component are mainly influenced by starch, not gluten.

In summary, a high score on PC1 indicates a high viscosity after gelatinization, which may be related to a low alpha-amylase activity. PC1 can be best assessed by the micro visco amylograph and the rapid visco analyser.

Water absorption and dough plasticity/elasticity at fixed water amount

PC2 describes water absorption and the dough plasticity/elasticity ratio at fixed water amount. The level of damaged starch, as assessed by the SDmatic (DST) and the SRC sodium carbonate value (SRC), has a strong influence on hydration, as assessed by the SRC water value (SRW), and the Mixolab and Farinograph hydration values (XHY, FWA). Similar results were obtained by Hammed et al. [10].

It is, furthermore, well understandable that the hydration is related to the plastic/elastic characteristic of the dough, as assessed with the AlveoLAB: the higher the level of water binding (by the starch but also by the gluten and the pentosans), the dryer the dough will be, which results in a lower extensibility (AVL, AVG), and a higher resistance to deformation (AVP), increasing the ratio P/L (AVR). This applies to the AlveoLAB measurement as it is performed with a fixed amount of water according to ICC 121, and is consistent with the results reported by Van Bockstaele et al. [21] and Li et al. [25]. The Extensograph measurement, on the contrary, is performed with a variable amount of water according to ICC 114/1, which probably explains why its values are not found on PC2.

The observed correlation of the RVA pasting temperature VAP with the SRC sodium carbonate (SRC) is in line the results of Barak et al. [2], Yu et al. [22], and Ma et al. [15] stating a correlation with the level of damaged starch, and can be interpreted as an earlier pasting in case of better hydration of the starch granules.

It is interesting that the SRC water and sodium carbonate values (SRW, SRC) are related to the PCD total charge after 4 and 24 min (P04, P24). Longin et al. [14] already observed this phenomenon on emmer flours. This suggests that a higher (negative) charge of the macroions in the dough leads to a higher water binding.

A high score on PC2 indicates a low water absorption, a low level of damaged starch, and a high dough extensibility at fixed water amount. PC2 can be well assessed by the measurement of the SRC values, the level of damaged starch, and the AlveoLAB extensibility.

Dough resistance at variable water amount

PC3 is related to dough resistance as measured with the Extensograph. PC3 shows that doughs that are standardized to a certain firmness (500 BU) by varying the amount of water in the Farinograph will still, after a resting time of 45, 90, and 135 min, have a different resistance to deformation (ER45, EM45, ER90, EM90, ER135, and EM135). As there is no relation of PC3 to the extensibility values, it is probable that the observed correlation with the ratio values (EY45, ED45, EY90, ED90, EY135, and ED135) is only related to the resistance values.

Remarkably, the amount of ascorbic acid has a high correlation with the ratio values (ED90, ED135, and EY135), which indicates that the addition of ascorbic acid favours elastic properties (resistance to deformation) at the expense of plastic properties (extensibility). This is consistent with the data of [17].

A high score on PC3 indicates a high resistance to deformation of dough at variable water amount. It can be well assessed with the Extensograph and can be, to a certain extent, influenced by the addition of ascorbic acid.

Dough strength at fixed water amount

PC4 describes dough strength, i.e., the combination of elastic and plastic properties, as measured by the AlveoLAB at fixed water amount. This property is related to the protein content and the GlutoPeak torque values (GPA, GPM). The latter is in line with the findings of Marti et al. [17], who developed correlation models to predict AlveoLAB values from GlutoPeak measurements.

A high score on PC4 indicates a high dough strength at fixed water amount. It can be well assessed with the AlveoLAB.

Other correlations

It is interesting that the AlveoLAB dough strength (AVW) is related to the Extensograph energy values (EE45, EE90, and EE135). Both values are based the same concept (combination of resistance and extensibility measured when stretching dough); despite the differences between both measuring systems (inflating a bubble of dough vs. stretching a dough piece) and the differences of water amount (fixed vs. variable), the values are still correlated—unlike the other values determined by AlveoLAB and Extensograph.

The correlations found between the values measured by the Farinograph and the Mixolab confirm previous findings, e.g., from [412]. The correlation between the hydration values and the GlutoPeak values suggests that a stronger gluten will bind more water. This in line with the data of Marti et al. [17] and Fu et al. [24] .

The absence of strong correlation between the level of gluten (wet and dry) and the measured functional values suggest that in our flour samples, the quality of the gluten was highly variable. This can be explained by our deliberate choice to use commercial flours from very various origins.

Conclusions

Our results highlighted the importance of the following properties in wheat flour analysis: starch gelatinization, hydration, dough resistance at variable water amount, and dough strength at fixed water amount. These results may be of interest for millers and bakers when it comes to selecting methods appropriate to describe quality in an effective yet comprehensive manner. It seems indeed advisable to use at least one method from each of the four groups defined by the principal components. Several options are available regarding the choice of instruments. On the contrary, it probably makes less sense to use two methods which results are highly correlated to one another. It is to be noted that, as flour processing companies will hardly invest in both an AlveoLAB and an Extensograph, the possibility to work with a single instrument at both fixed and variable water amount should be explored.

Similarly, we suggest that research teams may use our results to select the most appropriate setting of instruments to characterize their wheat flours—although this may result in a different selection of instruments than in processing companies, as the time of analysis may be less relevant.

The values assessed by the emerging methods SRC (as performed with the SRC-CHOPIN) and Brabender GlutoPeak show to have high correlations with established methods. This suggests that these methods do not assess new properties of flour, but rather represent an alternative to established methods—especially considering that they are faster and that they may be more cost effective.

As far as the baking performance is concerned, it is of course of interest to assess which of the investigated methods are mostly related to the properties of the final baked goods. Corresponding data will be published by our project team in a further article [23].

References

1. Aamodt A, Magnus EM, Faergestad EM (2003) Effect of flour quality, ascorbic acid, and DATEM on dough rheological parameters and hearth loaves characteristics. J Food Sci 68:2201–2210

2. Barak S, Mudgil D, Khatkar BS (2013) Relationship of gliadin and glutenin proteins with dough rheology, flour pasting and bread making performance of wheat varieties. LWT-Food Sci Technol 51:211–217

3. Cauvin S (2015) Technology of breadmaking. Springer, Heidelberg

4. Dabčević T, Hadnađev M, Pojić M (2009) Evaluation of the possibility to replace conventional rheological wheat flour quality control instruments with the new measurement tool—Mixolab. Agric Conspec Sci 74:169–174

5. Dalla Marta A, Grifoni D, Mancini M, Zipoli G, Orlandini S (2011) The influence of climate on durum wheat quality in Tuscany, Central Italy. Int J Biometeorol 55:87–96

6. Deffenbaugh LB, Walker CE (1989) Comparison of starch pasting properties in the Brabender Viscoamylograph and the Rapid Visco-Analyzer. Cereal Chem 66:493–499

7. El-Hady EA, El-Samahy SK, Brümmer JM (1999) Effect of oxidants, sodium-stearoyl-2-lactylate and their mixtures on rheological and baking properties of nonprefermented frozen doughs. LWT-Food Sci Technol 32:446–454

8. Fierens E, Helsmoortel L, Joye IJ, Courtin CM, Delcour JA (2015) Changes in wheat (Triticum aestivum L.) flour pasting characteristics as a result of storage and their underlying mechanisms. J Cereal Sci 65:81–87

9. Fremaux A (2015) Bake-off bakery markets in the EU. Gira, Paris

10. Fu BX, Wang K, Dupuis B (2017) Predicting water absorption of wheat flour using high shear-based GlutoPeak test. J Cereal Sci 76:116–121

11. Hammed AM, Ozsisli B, Ohm JB, Simsek S (2015) relationship between solvent retention capacity and protein molecular weight distribution, quality characteristics, and breadmaking functionality of hard red spring wheat flour. Cereal Chem 92:466–474

12. Huen J, Börsmann J, Matullat I, Böhm L, Stukenborg F, Heitmann M, Zannini E, Arendt EK (2017) Pilot scale investigation of the relationship between baked good properties and wheat flour analytical values. doi: 10.1007/s00217-017-2975-2

13. Koksel H, Kahraman K, Sanal T, Ozay DS, Dubat A (2009) Potential utilization of Mixolab for quality evaluation of bread wheat genotypes. Cereal Chem 86:522–526

14. Kweon M, Slade L, Levine H (2011) Solvent retention capacity (SRC) testing of wheat flour: principles and value in predicting flour functionality in different wheat-based food processes and in wheat breeding—a review. Cereal Chem 88(6):537–552

15. Li J, Hou GG, Chen Z, Chung A-L, Gehring K (2014) Studying the effects of whole-wheat flour on the rheological properties and the quality attributes of whole-wheat saltine cracker using SRC, alveograph, rheometer, and NMR technique. LWT Food Sci Technol 55(1):43–50

16. Longin F, Ringer L, Lösche K, Starck N, Römer P (2015) Development of germplasm and strategies for sustainable breeding of Emmer (Triticum dicoccum) for organic farming in Germany, BÖLN project report. http://orgprints.org/27958/1/27958-10OE059-V-uni-hohenheim-longin-2014-qualitaetszuechtung-emmer.pdf. Accessed 20 Apr 2017

17. Ma S, Li L, Wang XX, Zheng XL, Bian K, Bao QD (2016) Effect of mechanically damaged starch from wheat flour on the quality of frozen dough and steamed bread. Food Chem 202:120–124

18. Marti A, Augst E, Cox S, Koehler P (2015) Correlations between gluten aggregation properties defined by the GlutoPeak test and content of quality-related protein fractions of winter wheat flour. J Cereal Sci 66:89–95

19. Marti A, Ulrici A, Foca G, Quaglia L, Pagani MA (2015) Characterization of common wheat flours (Triticum aestivum L.) through multivariate analysis of conventional rheological parameters and gluten peak test indices. LWT-Food. Sci Technol 64(1):95–103

20. Peña RJ, Posadas-Romano R (2013) Potential use of the mixolab in wheat breading. In: Dubat A, Rosell CM, Gallagher E (eds) Mixolab: a new approach to rheology. AACC International Inc, St. Paul, pp 79–84

21. Schick J, Lösche K (2010) Characterization of milled grain products by particle charge detection. In: Keil HM (ed) Review "Bread", f2m food multimedia, Hamburg, pp 130–139

22. Semenov MA, Stratonovitch P, Alghabari F, Gooding MJ (2014) Adapting wheat in Europe for climate change. J Cereal Sci 59:245–256

23. Van Bockstaele F, De Leyn I, Eeckhout M, Dewettinck K (2008) Rheological properties of wheat flour dough and the relationship with bread volume. I. Creep-recovery measurements. Cereal Chem 85:753–761

24. Yu J, Wang S, Wang J, Li C, Xin Q, Huang W, Zhang Y, Zhonghu H, Wang S (2015) Effect of laboratory milling on properties of starches isolated from different flour millstreams of hard and soft wheat. Food Chem 172:504–514

Chapter 4 – Pilot scale investigation of the relationship between baked good properties and wheat flour analytical values

Abstract

Our objective was to assess the relationship between wheat flour analytical values and baked good properties based on equipment and processes near to those of mid-size bakery companies, including fermentation interruption. Thirty-four physical properties of rolls and breads were measured, and their correlation with 90 analytical properties of flour was investigated. In rolls, fermentation interruption led to a coarser crumb structure and a lower height/width ratio as compared to direct fermentation. This can be seen as a quality loss, which extent varied considerably depending on the flour. Our results suggest that using flours with a strong gluten network and high water absorption properties are most appropriate for fermentation interruption. Using a PLS regression, we identified three flour properties that had the highest overall influence on the final baked goods properties: flour hydration, dough stability, and elasticity/plasticity ratio.

Keywords

Bread properties, Wheat flour, Partial least-square regression, Interrupted fermentation

This chapter has been published:

Huen J, Börsmann J, Matullat I, Böhm L, Stukenborg F, Heitmann M, Zannini E, Arendt EK (2018).

Pilot scale investigation of the relationship between baked good properties and wheat flour analytical values. *European Food Research and Technology*, 244 (3), 481-490. doi:10.1007/s00217-017-2975-2

Introduction

In the food industry and more particularly in bakery companies, extensive quality and process data are measured and monitored: analytical values of raw materials, process parameters, and analytical values of final products. However, it is striking that these data are most of the time only compared with reference values (specifications) and seldom analysed in relation to one another. We believe that there is an unexploited potential in bakery companies to analyse their data for understanding the factors influencing the quality of final baked goods. To illustrate this approach, in the present study, we gathered analytical and process data within investigations performed at pilot scale and investigated their relationships. We thereby used commercial wheat flours from a variety of European mills, as well as equipment and processes typical for mid-size craft bakeries, including fermentation interruption (i.e., interrupting proofing by a freezing step to gain flexibility). Using statistical tools, we aimed at identifying analytical values most relevant for the quality of the final baked goods. Extensive work has already been performed in the past by numerous authors on the relationship between flour and bread properties, as summarized by Stojceska et al. [15]. However, most of the publications available are based on lab scale baking and flour from well-controlled origin. It was, therefore, an objective of this work to explore whether the relationships described in the literature also apply when using a baking approach close to business practice. The 37 flours used in this study have been analysed by a comprehensive set of methods, generating 90 analytical values for each, as described in our previous paper [10]. As could be expected, these 90 values are not independent from one another. Rather, a principal component analysis (PCA) showed that four underlying flour properties accounted for 64.8% of the total variance in the data set. These flour properties were identified as starch gelatinization properties, hydration properties, dough resistance at variable water amount, and dough strength at fixed water amount. The present article explores how the analytical values are related to the baked good properties.

Experimental

The baking trials were performed using the 37 commercial wheat flours from 14 mills located in seven European countries that we investigated analytically as described in our previous paper.

Baking process

All 37 flours were baked according to the following procedure (Tables 4.1 and 4.2): flour, water, salt (Suprasel fine, Suprasel, The Netherlands), dry yeast (Pante, Puratos, Burmingham, UK), and fat (Goldcup Palm, Vandemoortele, Herford, Germany); all tempered at 20 °C, were homogenised for 2 min at low speed and kneaded to a dough for 5 min at high speed in a kneader SP12 (Diosna, Osnabrück, Germany). Two different amounts of water were used in the experiments: a fixed amount of 56% (series A), and a variable amount (series B) assessed as water absorption (FWA) with the Farinograph (Brabender, Duisburg, Germany). After kneading and a resting time of 5 min, the dough was divided into three portions and further processed as described in Table 4.2, leading to rolls from direct fermentation, rolls from fermentation interruption, and toast breads. For the fermentation interruption, a time/temperature process typical for mid-sized bakery companies was applied, including a freezing step at -18 °C. The toast breads were baked with covers on the moulds, leading to breads of identical volumes. All baking trials were performed in duplicate on 2 different days. For each of the flours, all analyses and bakery trials were conducted within 2 weeks, to exclude as far as possible changes of flour properties during storage.

Table 4.1: Recipe

Ingredient	Percent of flour	Weight [g]
Wheat flour	100,0%	4453,0
Water	56,0% or variable	2494,0 or variable
Salt	2,2%	98,0
Dry yeast	2,0%	89,0
Palm fat	1,0%	44,5
Sum	**161,2% or variable**	**7178,5 or variable**

Table 4.2: Bakery process. The first 3 steps are common to the 3 portions.

Process step	Equipment	Parameters	Portion 1	Portion 2	Portion 3
Kneading	4 Diosna SP12	2 min low speed 5 min high speed			
Resting	-	5 min			
Portioning	-	(1) 2160 g + (2) 2160 g + (3) 2400 g			
Rounding and rolling out	Manually		●	●	
Roll pressing	Roll press	2160 g → 30 x 72 g	●	●	
Divide and round dough, put in mould	Baking mould with cover	4 x 600 g			●
Rising	MIWE GR	35 °C, 75% rh	70 min		85 min
Fermentation interruption	MIWE GVA	3:45 h at -18 °C 2:30 h rise -18 → -2 °C 4:45 h at -2 °C 7:00 h rise -2 °C → 18 °C 4:00 h at 18 °C		●	
Baking	MIWE roll in e+	220 °C, 20 min, 3 L steam at start			●
	MIWE condo	220 °C, 35 min, 400 mL steam in the last 30 s	●	●	
Cooling	-	1 hour at room temperature	●	●	●

Baked good characterization

After baking, the rolls and breads were allowed to cool down for 1 h at ambient conditions. A C-Cell (Calibre Control, Warrington, UK) was used to determine the cell (pore) density and the number of holes (hole size sensitivity: 4,5). For the rolls, a VolScan Profiler (Stable Micro Systems, Godalming, UK) was used to determine the specific volume and the height/width ratio. For the breads, a texture analyser (Stable Micro Systems) was used to determine the hardness and the elasticity of the crumb after 1 day of storage. A code was assigned to each value (Table 4.3).

Data processing

Pearson's correlation coefficients between flour analytical values and parameters describing the quality of the final baked goods, with the corresponding P values, as well as a PLS regression with four latent variables, were calculated using XLSTAT Version 2015 (Addinsoft, Paris, France).

Results and discussion

Table 4.3 gives an overview of the data obtained. Table 4.4 presents the correlation coefficients with an absolute value higher than 0.5 and $P < 0.001$. Table 4.5 shows the correlation coefficients of the investigated parameters with the first four latent components. Table 4.6, finally, shows the analytical values with the highest variable importance in the projection (VIP).

Influence of the amount of water

The water absorption determined by the farinograph was 58.1% in average, and only 3 flours out of 37 had a water absorption lower than 56.0%, the fixed water amount used in the A series. The average dough temperature was higher in the dryer and thus more viscous A series (DTA 23.2 vs. DTB 22.7 °C). The rolls of the B series had a significantly higher specific volume as well as a significantly lower height/width ratio and cell density than the rolls of the A series, both in direct and indirect fermentation (specific volume direct VDA 3.72 < VDB 3.82 L.kg^{-1}; indirect VIA 3.88 < VIB 4.02 L.kg^{-1}; height/width ratio direct VXA 0.65 > VXB 0.63; indirect VYA 0.58 > VYB 0.56; cell density direct CDDA 0.66 > CDDB 0.65 mm^{-2}; indirect CDIA 0.59 > CDIB 0.58 mm^{-2}). The significance of the differences was assessed by T tests with a threshold of

0.01. These results show how dough viscosity (here: influenced by the amount of water) impacts the quality of rolls: a higher hydration leads to a softer dough, allowing for a better expansion (higher specific volume and lower cell density) yet at the expense of the height/width ratio (flatter rolls). These effects are discussed more thoroughly under "Correlation with flour analytical data". Breads from the B series were significantly softer than in the A series (hardness THA 8.5 > THB 8.2 N). This can be related to the lower proportion of solid matter (in average) in the dough.

Table 4.3: Overview of baking results obtained with the 37 investigated flours (mean, min, max, relative standard deviation). R = number or repetitions. dl = dimensionless.

Method / Device	R	Output	Code	Mean	Min	Max	rsd	Unit
Dough temperature	2	Dough temperature (A)	DTA	23,2	21,8	24,7	2,8%	°C
		Dough temperature (B)	DTB	22,7	21,6	24,3	3,1%	°C
Stable Micro Systems VolScan Profiler	2 x 10	Specific volume of rolls – direct fermentation (A)	VDA	3,72	3,39	4,28	7,0%	L.kg^{-1}
		Specific volume of rolls – direct fermentation (B)	VDB	3,82	3,50	4,39	6,3%	L.kg^{-1}
		Specific volume of rolls – fermentation interruption (A)	VIA	3,88	3,35	4,69	8,4%	L.kg^{-1}
		Specific volume of rolls – fermentation interruption (B)	VIB	4,02	3,21	4,68	8,7%	L.kg^{-1}
		Specific volume of rolls – relative difference (A)	VQA	4,6%	-12,9%	25,3%	177%	
		Specific volume of rolls – relative difference (B)	VQB	5,4%	-17,8%	22,7%	176%	
		Height/Width of rolls – direct fermentation (A)	VXA	0,65	0,60	0,68	2,4%	- (dl)
		Height/Width of rolls – direct fermentation (B)	VXB	0,63	0,58	0,68	3,4%	- (dl)
		Height/Width of rolls – fermentation interruption (A)	VYA	0,58	0,46	0,66	6,0%	- (dl)
		Height/Width of rolls – fermentation interruption (B)	VYB	0,56	0,48	0,62	7,0%	- (dl)
		Height/Width of rolls – relative difference (A)	VRA	-10,5%	-23,8%	-0,6%	-41,2%	
		Height/Width of rolls – relative difference (B)	VRB	-10,8%	-19,3%	-1,6%	-42,5%	
Calibre Control C-Cell	2 x 10	Cell density of rolls – direct fermentation (A)	CDDA	0,66	0,56	0,75	6,8%	mm^{-2}
		Cell density of rolls – direct fermentation (B)	CDDB	0,65	0,55	0,70	5,5%	mm^{-2}
		Number of holes in rolls – direct fermentation (A)	CHDA	1,05	0,33	2,26	38,9%	- (dl)
		Number of holes in rolls – direct fermentation (B)	CHDB	0,92	0,28	1,54	35,1%	- (dl)
		Cell density of rolls – fermentation interruption (A)	CDIA	0,59	0,54	0,65	5,0%	mm^{-2}
		Cell density of rolls – fermentation interruption (B)	CDIB	0,58	0,52	0,61	3,6%	mm^{-2}
		Number of holes in rolls – fermentation interruption (A)	CHIA	1,48	0,74	2,29	25,5%	- (dl)
		Number of holes in rolls – fermentation interruption (B)	CHIB	1,54	0,66	2,25	25,5%	- (dl)
		Cell density of rolls – relative difference (A)	CDQA	10,3%	20,5%	-1,5%	-10,3%	
		Cell density of rolls – relative difference (B)	CDQB	-11,1%	-20,0%	-0,4%	-11,1%	
		Number of holes in rolls – relative difference (A)	CHQA	59,0%	-43,7%	294%	59,0%	
		Number of holes in rolls – relative difference (B)	CHQB	92,3%	-28,6%	486%	92,3%	
	2 x 8	Cell density in breads (A)	CDBA	0,52	0,46	0,59	5,1%	mm^{-2}
		Cell density in breads (B)	CDBB	0,53	0,46	0,63	5,9%	mm^{-2}
		Number of holes in breads (A)	CHBA	0,89	0,20	2,39	49,9%	- (dl)
		Number of holes in breads (B)	CHBB	0,86	0,33	1,95	45,0%	- (dl)
Texture Analyser	2 x 12	Hardness of bread (A)	THA	8,5	6,8	10,9	9,6%	N
		Hardness of bread (B)	THB	8,2	6,3	10,4	11,7%	N
		Elasticity of bread (A)	TEA	87,7	70,7	95,2	6,9%	% (dl)
		Elasticity of bread (B)	TEB	87,9	74,4	96,6	8,0%	% (dl)

Table 4.4: Correlation coefficients between flour analytical values and baked goods properties with absolute values higher than 0,5 and P < 0,001.

Dough T

		GPk	Fari.	AlveoLAB			
		GPS	FWA	AVP	AVL	AVG	AVR
Dough T	DTB	-0,55	-0,55	-0,53	0,56	0,55	-0,51

Height/Width

		AscA	Farinograph			Mixo.	AlveoLAB			SRC
		ASA	FST	FDS	FQV	XHY	AVP	AVW	AVI	SRL
Height/ Width	VXB	-0,53								
	VYA					0,60	0,55	0,68	0,55	0,60
	VRA		0,53	-0,53		0,56	0,55	0,69	0,57	0,58
	VRB		0,62	-0,59	0,64			0,65	0,59	

Volume

		Ash
		ASH
Volume	VQA	-0,55

Cell density

		GlutoPeak						Mixolab				Da.St.	Fari.	AlveoLAB			SRC-CHOPIN		Rapid Visco Analyser				
		GPT	GPM	GPB	GPA	GPS	GPG	X3C	XHY	X5D	AVP	DST-	FWA	AVL	AVG	AVR	SRL	SRC	VAM	VAY	VAV	VAS	VAP
Cell density	CDDA			0,62	0,55	0,55	0,55		0,62	-0,65	0,52		0,68	-0,53	-0,52				-0,54	-0,53	-0,56	-0,60	-0,52
	CDDB			0,63	0,56	0,55	0,55		0,64	-0,50	0,56		0,62	-0,55	-0,53								-0,56
	CDIA	-0,52		0,58		0,52			0,59	0,61	0,61	0,55	0,54	-0,66	-0,66	0,65		0,60					-0,52
	CDQB		-0,56	-0,60			-0,65	0,54	-0,58		-0,50		-0,62										
	CDBA																0,64						
	CDBB																0,56						

Holes

		Sedi.	GlutoPeak				µVisco Amylo.			Da.St.	Fari.
		SDV	GPB	GPA	GPG	GPS	MVB	MVC	MVX	DST	FWA
Holes	CHDA	0,64	0,54	0,50			-0,52		-0,51		0,54
	CHDB	0,50			0,55						
	CHIA		0,59	0,53						0,61	
	CHIB						0,55	0,56			
	CHQA	-0,57									
	CHQB										

		Mixolab		Alveo	RF4	SRC-CHOPIN			Rapid Visco Analy.		
		X3D	X5D	AVP	RGT	SRS	SRL	SRC	VAM	VAS	VAP
Holes	CHDA					0,52	0,56			-0,59	-0,57
	CHDB		-0,56	0,52			0,65	0,50			
	CHIA										
	CHIB										
	CHQA	0,61			-0,57		-0,51		0,57	0,56	
	CHQB										

Elasticity

		µVA	GMa	Farinograph			Mixolab			AlveoLAB		SRC Chopin		RVA
		MTA	GMW	FST	FDS	FQV	XHY	X2C	X4C	AVP	AVW	SRW	SRL	VAB
Elasticity	TEA	-0,58		0,60	-0,66	0,56	0,55	0,65		0,54	0,57	0,54	0,68	-0,66
	TEB	-0,54	0,54	0,64	-0,72	0,58	0,54	0,64	0,53	0,60	0,58	0,63	0,57	-0,69

Direct vs. indirect fermentation

At both water levels, fermentation interruption led to a significantly higher specific volume and number of holes, and a significantly lower height/width ratio and cell density than direct fermentation (increase in specific volume between direct and indirect VQA 4.6%, VQB 5.4%, increase in number of holes CHQA 59.0%, CHQB 92.3%, decrease of height/width ratio VRA −10.5%, VRB −10.8%, decrease of cell density CDQA −10.3%, and CDQB −11.1%). The higher specific volume as well as the lower cell density and height/width ratio may be explained in two ways: on one hand, gluten may be damaged by the freezing/thawing process [14], reducing its ability to resist deformation during proofing. On the other hand, freezing and thawing lead to a migration of water in the dough [1]. Water released by the melting of ice crystals may not be reabsorbed by starch, gluten, and pentosans, leading to a higher amount of unbound water after thawing [7], thus reducing viscosity. The higher number of holes may be created by ruptures of the dough film separating cells during freezing and thawing, as described by Gélinas et al. [9] and Lucas et al. [13].

Correlation with flour analytical data

Cell density

In rolls from direct and indirect fermentation, the observed cell density values show significant correlations with analytical values related to both gluten and starch (GlutoPeak torque GPA and GPB, aggregation energy GPG, AlveoLAB tenacity AVP, Farinograph, and Mixolab hydration FWA and XHY, damaged starch DST, SRC-CHOPIN calcium carbonate SRC, Rapid Visco Analyser pasting temperature VAP and viscosities VAM, VAY, VAV, and VAS). In bread, cell density (CDBA and CDBB) appears to be significantly related solely to the SRC lactic acid value SRL.

From the literature data [17], we expect the size of the cells and thus the cell density to be related (1) to the quantity of gas produced, (2) to the capacity of the dough to hold the gas, and (3) to the resistance of the dough to deformation. The AlveoLAB tenacity AVP clearly measures the property (3). The GlutoPeak torque values GPA and GPB measure the strength of the gluten network, which can be related both to the properties (2) and (3). The correlation of cell density with the hydration values can be related to dough firmness (property 3): As hydration is defined according to a reference firmness, a higher measured hydration implies in a higher firmness in the A series (constant water amount).

The correlation of the cell density with the pasting temperature VAP may be interpreted in the following way: during baking, we expect the size of the cells to increase with temperature according to the expansion of gas, until the structure is finally stabilised by starch gelatinization [17]. Therefore, if gelatinization occurs at a higher temperature, a higher cell size can be reached.

The correlation with the RVA viscosity values VAM, VAY, VAV, and VAS can be explained by a lower expansion in the case of a higher viscosity. It can thus be related to dough resistance to deformation (property 3 as stated above) at higher temperature.

In the case of bread, cell density must be interpreted differently than in the case of rolls: as we used covers on the moulds, the volumes of all breads were identical and the gas expansion in the cells was limited. The correlation of the cell density with the SRC lactic acid value can be explained by a higher glutenin functionality [11], creating a denser protein network in the dough.

Holes

Most analytical factors related to cell density are related in the same way to the number of holes in rolls: GlutoPeak GPB, GPA, Farinograph FWA, damaged starch DST, Mixolab XHY, X5D, AlveoLAB tenacity AVP, SRC calcium carbonate, Rapid Visco Analyser VAS, and VAP. This seems contradictory in the first moment, as one would expect from a more compact crumb that it has less holes. In addition, the numbers of holes in rolls from direct fermentation CHDA and CHDB are positively correlated to the sedimentation value SDV, suggesting, like the GlutoPeak values and the AlveoLAB tenacity, that doughs with a stronger gluten network will have more holes. These observations suggest that holes are not large cells but are created, at least partly, by a different mechanism. A possible interpretation is that some holes are due to the incorporation of air pouches during the forming of dough pieces in the dough press, which would occur moreoften with a firmer dough that presents more resistance to deformation.

Height/width ratio of rolls

From the literature, little is known about the height/width ratio of rolls, as most experimental baking trials are performed in moulds, e.g., according to ICC 131. In the manufacturing process of rolls, it is important that the dough has a viscosity sufficient for avoiding spreading out, which would result in too flat baked goods.

In indirect fermentation, the positive correlation of the height/width ratio VYA with dough hydration XHY, SRC lactic acid SRL, AlveoLAB tenacity AVP, strength AVW, and Index AVI can be explained in terms of viscosity: a flour that absorbs more water and/or has a stronger gluten network will lead to a more viscous dough. In the case of direct fermentation (VXB), the observed negative correlation with the level of ascorbic acid ASA may seem paradoxical, as ascorbic acid should strengthen the gluten network. One explanation could be that the improving effect of ascorbic acid on the gas retention capacity [4] may result in a higher volume of the dough pieces (as observed by El Hady et al. [6]), itself inducing a lower overall viscosity of the fermented dough.

The fact that the Farinograph stability and dough softening values FST and FDS as well as the AlveoLAB strength and index values AVW and AVI seem to be helpful for identifying flours that are most appropriate for fermentation interruption (minimized height/width difference VRA, VRB) is remarkable, since these instruments do not simulate a freezing process. This finding is in line with data reported previously by Lu and Grant [12], Boehm et al. [2], and Gelinas et al. [9] and Frauenlob et al. [8], and can be related to gluten quality.

Specific volume

In our trials, no single analytical value had an absolute correlation coefficient of more than 0.5 and a *P* value below 0.001 with the specific volume values VDA, VDB, VIA, and VIB. On that point, our results differ from those of numerous other authors, e.g. van Bockstaele et al. [16] who found a correlation coefficient of 0.75 with the protein content, 0.74 with the sedimentation value, and 0.75 with the water absorption; Dobraszczyk and Salmanowicz [5] who report a correlation coefficient of 0.74 with the wet gluten content and 0.85 with the sedimentation value, or Bouachra et al. [3] who obtained correlation coefficients between 0.50 and 0.67 with GlutoPeak torque values. The absence of strong correlation of the specific volume with any single analytical value may be related both to the complexity of the pilot production process of the rolls and to the heterogeneity of the commercial flours used (37 flours from 14 mills in seven countries).

The only effect we observed, i.e., the influence of the ash content ASH on the volume increase due to fermentation interruption VQA, could be related to osmotic effects: with a higher ash content, the formation of ice crystals may be reduced, reducing the damage to the microstructure.

Elasticity of bread

Analytical parameters related to gluten quantity and quality (wet gluten content GMW, SRC lactic acid value SRL) as well as dough stability (Farinograph stability FST, quality value FQV, AlveoLAB tenacity AVP and strength AVW) are positively related to crumb elasticity TEA and TEB: this suggests that there is a continuum of the gluten properties before and after baking, a stronger gluten network allowing for a higher elasticity. This effect has, to our knowledge, not been reported yet by other authors.

Latent variables

The latent variables can be interpreted as hydration properties (LV1—based on the high correlation with Mixolab, Farinograph, and Solvent Retention Capacity hydration values XHY, FWA, and SRW), dough stability (LV2—based on the high correlation with the Farinograph and the Mixolab stability values FDS and XST), and dough elasticity/plasticity ratio at variable hydration (LV3—based on the high correlation with EDY45, EY90, and EY135), with LV4 being not clearly interpretable. These latent variables are linked, respectively, to crumb coarseness of rolls (LV1), crumb elasticity of bread (LV2), and negatively to the specific volume of rolls and the cell density of breads (LV3).

Referring to our previous paper describing the analytical properties of the 37 investigated wheat flours [10], LV1 is strongly correlated to PC2: the Pearson's correlation coefficient between both is $r = 0.86$. The main parameters with high values on both LV1 and PC2 are the Mixolab hydration XHY, the AlveoLAB AVP, AVL and AVR, the SRC-CHOPIN SRC and SRW, the damaged starch content DST, and the temperature of gelatinization VAP. LV2 is strongly correlated to PC1 ($r = 0.81$), the with main common parameters being Mixolab XST and X4C, Micro Visco Amylograph

Table 4.5: Correlation coefficients between the measured properties and the 4 latent variables (threshold 0,5).

	LV1		LV2		LV3		LV4
XHY	-0,93	FDS	-0,88	EX135	-0,76	VAB	-0,61
AVP	-0,90	FQV	0,88	EY90	0,70	MTA	-0,55
FWA	-0,87	FST	0,87	EY135	0,67	ASH	0,52
AVR	-0,79	PRT	0,75	ASA	0,66	SRW	0,50
SRC	-0,76	XST	0,75	EY45	0,65	RGS	-0,50
AVL	0,76	MTB	0,75	EX90	-0,64		
GPS	-0,75	MVD	0,74	ED90	0,60		
AVG	0,75	X2C	0,73	EX45	-0,60		
SRW	-0,74	MVC	0,73	ED135	0,59		
GPA	-0,73	X1T	0,73	ER135	0,57		
GPG	-0,73	VAY	0,72	ED45	0,57		
VAP	0,71	AVI	0,72	ER90	0,55		
GPB	-0,70	VAV	0,71	RD1	0,52		
X5D	0,66	MVE	0,70	ER45	0,51		
RDR	0,66	MVB	0,70	SDV	-0,51		
GPT	0,66	GMD	0,69	PH	0,50		
DST	-0,64	VAT	0,69				
X3T	0,64	AVW	0,67				
SRS	-0,63	EE135	0,65				
X3D	0,63	VAS	0,64				
SRL	-0,62	X4C	0,63				
GPM	-0,61	VAM	0,62				
SDV	-0,61	FAN	0,62				
VAM	0,61	VAX	-0,62				
X4D	0,59	EE45	0,61				
AVW	-0,58	MVY	0,61				
X3C	0,58	EE90	0,57				
VAS	0,56	EM135	0,54				
MTA	0,55	X4D	0,54				
RGT	-0,55	M4T	-0,54				
P04	0,55	ER45	0,51				
VAV	0,52	EM90	0,51				
P24	0,52	X1C	0,50				
MVX	0,51						
VAB	0,50						

	LV1		LV2		LV3		LV4
CDDA	-0,68	TEB	0,59	VIA	-0,60	VDB	0,64
CHDA	-0,67	VRB	0,55	CDBB	-0,59	THB	-0,61
CDDB	-0,66	TEA	0,52			VDA	0,60
CHQA	0,64					THA	-0,59
CDIA	-0,63						
CDQB	0,59						
DTB	0,58						
VYA	-0,56						
CHDB	-0,56						
VRA	-0,52						

MVC, MVD, MVE, and MTB, Rapid Visco Analyser VAM, VAY, VAV, and VAT values, and the falling number FAN).

Finally, LV3 is correlated to PC3 ($r = 0.74$), mainly via the Extensograph EY45, EY90, ED45, ED90, ED135, and ER135 values, as well as the ascorbic acid content ASA. Hence, the baking trials confirmed the importance of the first three principal components and their related flour parameters for baking companies.

Variable importance in the projection

The variables with the highest importance in the projection include, on one hand, the level of ascorbic acid, basic values like ash content, moisture, sedimentation value, protein content and falling number, and values determined by the use of specific commercial instruments: SRC-CHOPIN, GlutoPeak, Mixolab, Rapid Visco Analyser, Rheo F4, Farinograph, AlveoLAB, Glutomatic, Micro Visco Amylograph, and PCD. It is remarkable that the emerging methods SRC and GlutoPeak reached high scores on the VIP scale.

Most of the analytical values with the highest VIP have important direct correlations with the main properties of the final products, as discussed above. This is especially the case for the SRC lactic acid and water values SRL and SRW, the GlutoPeak torque values GPA and GPB, the Mixolab hydration and torque values XHY, X3D and X5D, the RVA viscosity values VAB, VAS, and gelatinization temperature VAP, the AlveoLAB strength AVW, tenacity AVP and index AVI, the level of ascorbic acid ASA, and the sedimentation value SDV. Next to these parameters, others which appear more rarely or not at all in the correlation tables still reached high scores. This suggests that these parameters, although being only weakly correlated with product properties taken individually, still have a significant contribution to the total baking result and should be monitored and incorporated in modelling activities. This is the case of the ash content ASH, the moisture MOI, and the Rheofermentometer gaseous release Tx RGX.

Generally speaking, our results are in line with findings made by other authors who baked at lab scale. However, the correlation coefficients found in our work are generally lower. This suggests that the variety of flours and the complexity of the baking process used in our trials are responsible for a higher variability. Stojceska and Butler [15] in their review on the relationship between rheological properties and baking performance already came to the conclusion that the correlations observed are highly dependent on the testing condition.

Table 4.6: Variable importance in the projection (VIP), threshold 1,0.

	Variable	VIP		Variable	VIP		Variable	VIP
Ascorbic acid	ASA	1,56		AVP	1,24		AVR	1,08
SRC Chopin	SRL	1,50		X5D	1,21		MVX	1,07
Ash content	ASH	1,44	Micro Visco Amylograph	MTA	1,20		AVI	1,07
GlutoPeak	GPB	1,42	Damaged starch	DST	1,20	PCD	P24	1,05
	GPG	1,40		FQV	1,18		AVL	1,05
Mixolab	X3D	1,39		FST	1,18		P04	1,04
Moisture	MOI	1,38		FDS	1,16		AVG	1,03
Rapid Visco Analyser	VAB	1,37		SRC	1,15		SRS	1,03
	GPA	1,37	Glutomatic	GMI	1,15		GPS	1,02
	XHY	1,34		PH	1,13		VAP	1,01
Rheo F4	RGX	1,30		RG1	1,12		X3T	1,01
	SRW	1,30		GPM	1,12	Falling number	FAN	1,01
Farinograph	FWA	1,30		VAS	1,12		X3C	1,01
AlveoLAB	AVW	1,28	Protein content	PRT	1,10		RGC	1,00
Sedimentation value	SDV	1,26						

Conclusions

The results of the study confirm the complexity of the investigated relationships, as no single flour analytical value has a very strong correlation with the properties of the final baked goods. Several effects described by other authors at lab scale were confirmed in our trials at pilot scale on rolls and toast breads. When selecting analytical parameters appropriate for describing the quality of wheat flour, we suggest to set a priority on methods highly correlated with the three first latent values (themselves related to the three first principal components from our previous paper), to include methods that reached high VIP scores, and methods that are highly correlated with the properties of interest in the specific application considered.

As far as the height/width ratio, the cell density, and the number of holes in rolls as well as the elasticity of toast bread are concerned, our results suggest that the AlveoLAB, the GlutoPeak, the SRC-CHOPIN, the Mixolab, the Farinograph, and the Rapid Visco

Analyser measurements provide valuable information. For identifying flours appropriate for fermentation interruption, the Farinograph and AlveoLAB seem to be most helpful: selecting flours with high hydration, high stability, high tenacity, and high strength values will help both improving cell density and reducing the specific loss of height/width due to the freezing/thawing process.

In conclusion, and coming back to the initial idea, we believe that it would be advantageous for bakery companies to develop their own data mining of flour, process, and end product quality values. This would enable them to better understand the flour properties relevant for their applications. By including comprehensive data acquired over the years, powerful prediction models could be developed.

References

1. Baier-Schenk A, Handschin S, Conde-Petit B (2005) Ice in prefermented frozen bread dough—an investigation based on calorimetry and microscopy. Cereal Chem 82:251–255

2. Boehm DJ, Berzonsky WA, Bhattacharya M (2004) Influence of nitrogen fertilizer treatments on spring wheat (Triticum aestivum L.) flour characteristics and effect on fresh and frozen dough quality. Cereal Chem 81:51–54

3. Bouachra S, Begemann J, Aarab L, Hüsken A (2017) Prediction of bread wheat baking quality using an optimized GlutoPeak®-Test method. J Cereal Sci 76:8–16

4. Cauvain SP, Young LS (2001) Baking problems solved. Woodhead Publishing, Cambridge

5. Dobraszczyk BJ, Salmanowicz BP (2008) Comparison of predictions of baking volume using large deformation rheological properties. J Cereal Sci 47:292–301

6. El-Hady EA, El-Samahy SK, Brümmer JM (1999) Effect of oxidants, sodium-stearoyl-2-lactylate and their mixtures on rheological and baking properties of nonprefermented frozen doughs. LWT Food Sci Technol 32:446–454

7. Esselink FJ, van Aalst H, Maliepaard M, van Duynhoven PM (2003) Long-term storage effect in frozen dough by spectroscopy and microscopy. Cereal Chem 80:396–403

8. Frauenlob J, Moriano M, Innerkofler U, D'Amico S, Lucisano M, Schoenlechner R (2017) Effect of physicochemical and rheological wheat flour properties on quality parameters of bread made from pre-fermented frozen dough. J Cereal Sci 77:58–65

9. Gélinas P, McKinnon CM, Lukow OM, Townley-Smith F (1996) Rapid evaluation of frozen and fresh dough involving stress conditions. Cereal Chem 73:767–769Google Scholar

10. Huen J, Börsmann J, Matullat I, Böhm L, Stukenborg F, Heitmann M, Zannini E, Arendt EK (2017) Wheat flour quality evaluation from the baker's perspective: comparative assessment of 18 analytical methods. Eur Food Res Technol. doi:10.1007/s00217-017-2974-3 Google Scholar

11. Kweon M, Slade L, Levine H (2011) Solvent retention capacity (SRC) testing of wheat flour: principles and value in predicting flour functionality in different wheat-based food processes and in wheat breeding—a review. Cereal Chem 88(6):537–552

12. Lu W, Grant LA (1999) Effects of prolonged storage at freezing temperatures on starch and baking quality of frozen doughs. Cereal Chem 76:656–662

13. Lucas T, Grenier D, Bornert M, Challois S, Quellec S (2010) Bubble growth and collapse in pre-fermented doughs during freezing, thawing and final proving. Food Res Int 43:1041–1048

14. Naito S, Fukami S, Mizokami Y, Ishida N, Takano H, Koizumi M, Kano H (2004) Effect of freeze-thaw cycles on the gluten fibrils and crumb grain structures of breads made from frozen doughs. Cereal Chem 81:80–86

15. Stojceska V, Butler F (2012) Investigation of reported correlation coefficients between rheological properties of the wheat bread doughs and baking performance of the corresponding wheat flours. Trends Food Sci Technol 24:13–18

16. Van Bockstaele F, De Leyn I, Eeckhout M, Dewettinck K (2008) Rheological properties of wheat flour dough and the relationship with bread volume. I. Creep-recovery measurements. Cereal Chem 85:753–761

17. Zhang L, Lucas T, Doursat C, Flick D, Wagner M (2007) Effects of crust constraints on bread expansion and CO2 release. J Food Eng 80:1302–1

Chapter 5 – Concluding remarks

Microstructure of frozen dough

The work performed in Chapter 2 shows that Raman microscopy allows for identifying ice, liquid water, starch, proteins and yeast cells in a frozen bread dough, and for imaging their respective spatial distribution within the samples. The images obtained in the measurements performed do not show single ice crystals but rather small ice blocks of 1 - 10 μm in diameter as well as a continuous three-dimensional ice crystal network. The boundaries between the individual ice crystals are not identifiable. It is unclear whether the observed ice blocks consist of one or several crystals each, but in any case, in can be assumed that the maximum ice crystal size in the observed sections of our frozen dough samples was 10 μm (maximal size of the observed ice blocks).

In this regard, the structure of ice in frozen dough is different from the one reported by other authors for frozen aqueous solutions or frozen animal and vegetal tissues (Lösche et al., 2014, Martino et al., 1998, Delgado et al., 2005, Otero et al., 2000, Sun, 2003, Do et al., 2004).

A possible interpretation of these observations is that large crystals can be formed only if enough space and water are available. In an aqueous solution, no solid structures are present which may mechanically hinder crystal growth, except other crystals. This is why the crystal size obtained during freezing is mainly dependent on the rate of nucleation, which is itself related to the the temperature gradient / the degree of supercooling.

In animal and vegetal tissues, the cytoplasm represents an aqueous solution. The only physical boundaries to crystal growth are the limits of the single cells (membranes) as well as the organelles, the latter being mobile within the cytoplasm.

In contrast, in bread dough, the three-dimentional gluten network and the embedded starch granules represent physical boundaries to ice crystal growth. In addition, water molecules involved in the hydration of gluten, starch and non-starch polysaccharides may not be available for crystal growth.

This interpretation suggests that in bread dough, the speed of freezing may have a more limited impact on ice crystal structure than in other foods, as the formation of large crystals is prevented by the structure of the dough itself. In the same way, the addition of ice structuring proteins to a dough was shown to have no observable effect on the ice crystal structure – as opposed to their application to aqueous solutions or very diluted dough models (Lösche et al., 2014).

It should be noted that, in the context of frozen foods, the concept of "ice crystal network" is not very common and has been mentioned only by few authors so far (Goff and Hartel, 2013, Locas et al., 2004, Lasalle et al., 2012, Windmoser et al., 2004, Zhong et al., 2015). An analogy can be made with the concept of "fat crystal network" in fat-based foods, as described by Deman and Beers (1987) and Marangoni (2004).

The fact that no or little ice is found within the starch granules (the Raman "starch" spectra measured within the samples are free from bands characteristics for ice) suggests that there is only little interaction between ice crystals and starch, and therefore limited possibility of starch damage attributable to ice. In contrast, the gluten network and the ice crystal network seem to be embedded in one another, giving more opportunity for interactions – although the nature of these interactions is still unclear. This represents the situation in freshly frozen dough, as investigated in Chapter 2.

During storage time, it is conceivable that ice crystals may slowly grow at the expense of proteins and / or starch structures, those being (further) damaged. Raman measurements of dough pieces after 6 months of frozen storage suggest this (Figure 5.3). Further investigations are necessary to develop a better understanding of these processes. However, it may be questioned whether dough storage times of several weeks or months are still relevant in regard of the current industry practice. From an industry point of view, it might be more pertinent to investigate the microstructure of frozen bread and its possible changes over long-term storage time.

Further possible developments and refinements of the microscopy method

Concerning the microscopy technique, it would be helpful to improve both speed and resolution. In the measurements of Chapter 2, integration time was one second at each point, leading to measurement times of approximately 12 hours per 100 x 100 µm image. The lateral resolution was around one μm while the vertical resolution was around two μm.

Measurement time may be reduced by the use of more sensitive detectors like EEMCD (electron multiplying charge-coupled devices), which allows measurement times below 1 ms in some applications, depending on the Raman yield of the substance investigated. Increasing resolution is possible by the use of better optical components (resolutions down to 200 nm laterally) or special techniques like Near-field Optical Microscopy (resolutions down to 100 nm laterally) (Toporski et al., 2018).

These improvements would enable investigating more samples in less time and imaging smaller structures, allowing for a better understanding of the ice crystal structure and its possible interactions with the other dough components.

From the point of view of sample preparation, in the measurements of Chapter 2, dough was frozen on the microscope slide. In order to investigate products frozen in real-life bakery processes, appropriate sample preparation techniques are necessary. Based on work conducted in the ice laboratory of the Alfred-Wegener Institut Bremerhaven, we propose a preparation technique involving sawing with a band saw followed by microtome dragging, as it is practiced with ice kernels in polar research (Figure 5.1 and 5.2). As soon as a smooth surface is obtained with the microtome, the sample should be protected with a microscope cover, as ice would otherwise sublimate fast in contact with the cold air. It has been shown that with samples prepared in this way, some dough structures are recognisable in reflected light microscopy. Thus, in this case, a faster alternative to Raman imaging as described in Chapter 2 consists of making point or line Raman measurements on the structures imaged in the reflected light modus (Figure 5.3). Based on the Raman spectra, the chemical nature of the observed structured can be assessed.

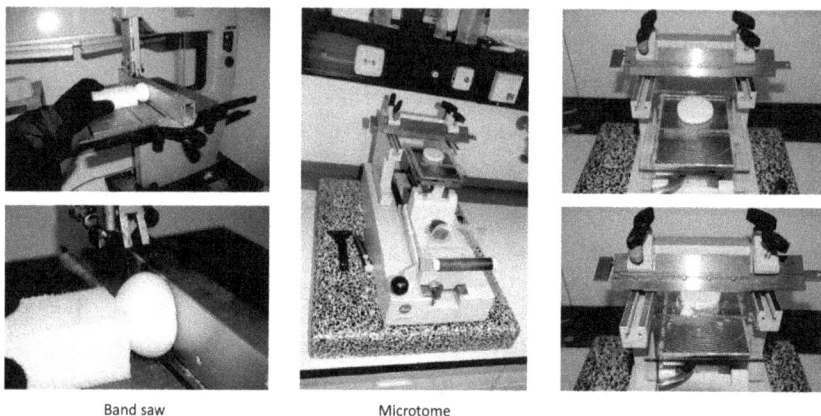

Band saw   Microtome

Figure 5.1: Sample preparation (frozen dough piece) with a band saw and a microtome in the ice laboratory at -15°C.

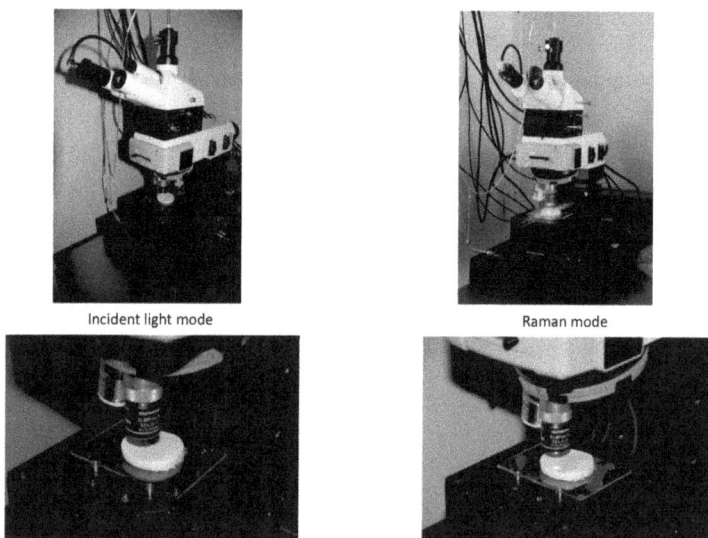

Incident light mode   Raman mode

Figure 5.2: Measurement of the frozen dough piece in the incident light and the Raman mode at -15°C.

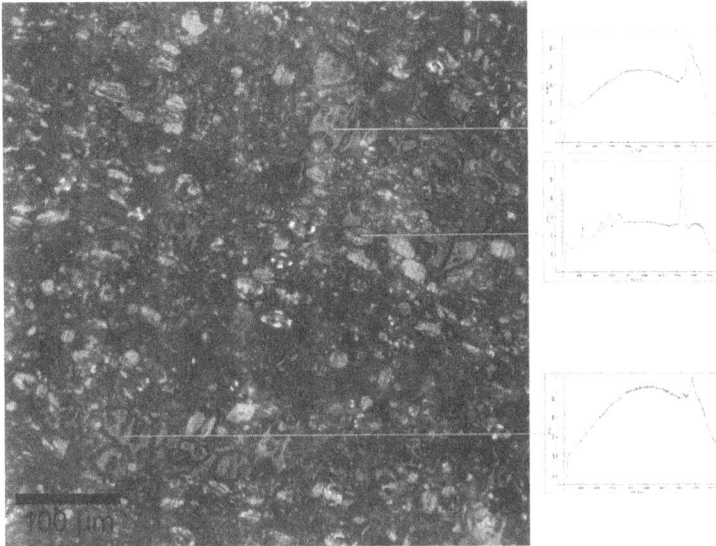

Figure 5.3: Microscope picture of a frozen dough piece (center part) in the incident light mode after 6 months of frozen storage at -18°C (measurement performed at -15°C).

Further potential applications of the (cryo) Raman microscopy in food science

With its unique capability of ice imaging, cryo Raman microscopy has the potential to be a powerful tool for investigating the microstructure of frozen food products. The only prerequisite is a plane surface, which can be obtained by microtome dragging. No further treatment of the sample is necessary, which could modify its properties, and a chemical characterization of the main sample structures is possible. The application to frozen cream with and without the addition of ISP has been published by our team (Roeder et al., 2017).

Raman microscopy may of course also be used for non-frozen food products. In all cases, it should be ensured that the product is stable enough during the measurement time. The weakness of the Raman signal is inherent to the method and represents its main drawback (Toporski et al., 2018).

It should be noted that Raman microscopy is well suited to show the microstructure of products with heterogeneous composition, involving substances with well

Chapter 5

distinguishable Raman spectra, due to greatly differing molecular structures. If the objective is to differentiate between molecules with similar Raman spectra like different protein types, the measurement may be more difficult (involve higher integration times) or not be possible at all.

Impact of fermentation interruption on the quality of rolls

The fermentation interruption process as described in Chapter 4 involves only a short-term storage at -18 °C (3:45 h). The process conditions were chosen based on typical business practice of German subsidiary bakeries. Core temperature measurements showed that this freezing time is sufficient to reach a core temperature of -15 °C, which, based on DSC measurements, allows the assumption that ice crystals are formed.

In our investigations, the fermentation interruption process had a significant impact on the quality of the rolls, as compared to direct fermentation. There was no loss of specific volume but, on the contrary, an increase (VQA 4,6 %, VQB 5,4 % in average). This suggests that under the processing conditions we chose, no particular lack of gas production due to yeast damage occured. Quality loss was observable in regard to height / width ratio (VRA -10.5 %, VRB -10.8 % in average), cell density (CDQA -10.3 %, CDQB -11.1 % in average) and number of holes (CHQA 59.0 %, CHQB 92.3 %), which means that the rolls were flatter and had a coarser structure.

These results confirm findings reported by subsidiary bakeries and explain why these companies may be willing to use specific bakery improvers to overcome the quality concerns observed.

Selecting analytical methods for flour specification

The results presented in Chapter 4 also show that the quality concern described above do not apply to all flours in the same way. This leads to the assumption that some flours are more appropriate than others for the fermentation interruption process and to the question how such flours may be selected and specified.

Given the importance of freezing processes in modern bakery production, it is striking that among the methods used in the supply chain to characterize flours, none directly assesses freezing stability. One objective of this thesis was to investigate whether the existing methods give indications on the appropriateness of flours for fermentation interruption.

In a first step, the work presented in Chapter 3 aimed at better understanding the information provided by the different flour analytical methods used across Europe. The assumption was that the 90 analytical values provided by 18 analytical methods are not independent from each other. When defining an analytical setup for flours, it makes sense not to select methods generating similar information. A PCA performed on our dataset based on 37 European flours showed that 64.8 % of the variance can be explained by four principal components. These components were interpreted as starch gelatinization properties, hydration properties, dough resistance at variable water amount and dough strength at fixed water amount. The data analysis showed which of the analytical values are particularly correlated with these four principal components.

In the next step, the aim was to assess the correlation of the analytical values with the bread properties. Numerous highly significant correlations were found (at a threshold of P < 0.001), but the correlation coefficients were relatively low (< 0.75). This reflects the multidimensional character of the investigated relationships – all properties of interest in the bread are influenced by a combination of factors. Several effects we observed are in line with the results of other authors (Bouachra et al., 2017, Dobraszszcyk and Salmanowicz, 2008, Frauenlob et al., 2017, Stojceska and Botler, 2012, Van Bockstaele et al., 2008). However, the correlation coefficients we found were generally lower. This may be explained by our choice to work on a flour selection representing the diversity of flours available on the European market, which introduces more variability to molecular composition than in the case of a national flour selection, or even a selection of flours based on controlled wheat varieties and growing conditions, as was the case in other studies.

Flour analytical properties found to have a relationship with the quality loss due to fermentation interruption are:

- in the case of the height / with ratio: Farinograph stability, dough softening and quality value, AlveoLAB tenacity, baking strength and elasticity index, SRC lactic acid value, Mixolab hydration.

- in the case of cell density: Glutopeak maximum torque, torque before and after maximum, aggregation energy, AlveoLAB tenacity, Farinograph water absorption, Mixolab hydration and C3 dough temperature.

- In the case of the number of holes: Sedimentation value, SRC lactic acid value, Mixolab hydration, Micro visco amylograph temperature of max viscosity, viscosity at start of holding and viscosity difference B-D, C3 and C5 torque, Rapid Visco Analyser Peak and Setback viscosity, Rheofermentometer gaseous release V_t,.

This suggests that analytical values related to a strong gluten network and to a high level of water absorption are related to better results in fermentation interruption. It should be noted that not the level of gluten seems to be the key, but rather its rheological properties. A high level of water absorption may be associated with a low level of ice in the frozen dough. In such a case, the components of flour responsible for water absorption (starch, gluten, non-starch polysaccharides) would act as natural cryoprotectants. This interpretation is confirmed by the fact that the freezing damage was less pronounced in the series (A) of the trials, in which a lower amount of water was used in almost all cases. In addition, the pasting properties of starch obviously play a role in hole formation after the freezing/thawing process.

Interpretation in the light of the microscopy results

The level of ice created seems to play a role on the importance of the quality loss due to freezing. It therefore appears that the quality loss is not (only) the effect of the temperature decrease itself, but is directly related to ice crystal formation. Microscope observations suggested that a high degree of interaction between the gluten network and the ice crystal network is possible. In contrast, little or no interaction between ice crystals and starch granules was observed.

Potential utilization of the presented results

The findings of Chapter 3 and 4 may help bakers and millers agreeing on flour specifications that deliver the most useful information based on the amount of analytical work practicable in daily business. In particular, analytical methods were identified that provide information on performance in fermentation interruption, as applied by subsidiary bakers. Better defining flour qualities has, in turn, the potential to reduce the use of additives and enzymes, satisfying current market demand.

The ability to deliver flours matching particular specifications depends on the availability of wheat that fulfils these requirements, either directly or in blends. Therefore, it is important that the qualities required by the bakeries and the corresponding volumes needed are known in the whole supply chain. Spreading this knowledge will enable wheat breeders, farmers and cereal traders to improve their products and processes in order to deliver the required qualities in a targeted way.

The findings of Chapter 2 on the use of Raman microscopy may be used by researchers for further microscopy investigations of frozen dough, frozen bread, and other frozen and unfrozen foods to better understand phenomena occurring at the microscopic level and their importance for functionality.

Outlook / potential for further research

While the results presented may help specifying flours appropriate for fermentation interruption, and confirmed the importance of ice formation in freeze damage of bread dough as well as the key role of gluten quality for minimizing those damages, the underlying mechanisms are still unclear. More insights may be obtained by further microscopic studies and by performing a molecular characterization of gluten and starch before freezing and after thawing, providing evidence of transformations occurring at the molecular level. In addition, it would be helpful to determine which gluten molecular compositions are particularly appropriate for fermentation interruption processes.

The fact that nowadays, industrial bakeries freeze bread rather than dough poses in a similar way the question of possible damages occurring during this application and of their causes at the molecular level. This could be the object of future investigations.

Outlook / potential for further industry developments

The food industry works on the basis of natural raw materials, which undergo quality variations. Industrial processing, however, is in the most cases based on fixed process conditions. Finding structured ways to react to raw material quality variations in production is a challenge for the future. Developing a better fundamental understanding of the product properties and their relationship to molecular structures on the one hand and learning better from production experience, e.g. based on data management tools and machine learning systems, are possible approaches.

Communicating requirements across supply chains is another challenge. Companies most often have a deep understanding of their own processes and of their interfaces with their direct suppliers and customers, but lack knowledge of the other steps of the supply chain. Spreading knowledge across the supply chains will be a key to serve changing markets.

References

Bouachra S, Begemann J, Aarab L, Hüsken A (2017). Prediction of bread wheat baking quality using an optimized GlutoPeak®-Test method. Journal of Cereal Science 76, 8-16.

Dobraszczyk BJ, Salmanowicz BP (2008). Comparison of predictions of baking volume using large deformation rheological properties. Journal of Cereal Science 47, 292-301.

Delgado AE, Rubiolo AC (2005). Microstructural changes in strawberry after freezing and thawing processes. LWT - Food Science and Technology 38, 135-142.

Deman JM, Beers AM (1987). Fat crystal networks: structure and rheological properties. Journal of Texture Studies 18, 303-318.

Do GS, Sagara Y, Tabata M, Kudoh KI, Higuchi T (2004). Three-dimensional measurement of ice crystals in frozen beef with a micro-slicer image processing system. International Journal of Refrigeration, 27, 184-190.

Frauenlob J, Moriano ME, Innerkofler U, D'Amico S, Lucisano M, Schoenlechner R (2017). Effect of physicochemical and empirical rheological wheat flour properties on quality parameters of bread made from pre-fermented frozen dough. Journal of Cereal Science 77, 58-65.

Goff IID, Hartel RW (2013). Ice cream structure. In: Goff HD, Hartel RW (Ed.), Ice cream, Springer US, 313-352.

Lucas T, Mariette F, Dominiawsyk S, Le Ray D (2004). Water, ice and sucrose behavior in frozen sucrose–protein solutions as studied by 1H NMR. Food Chemistry 84, 77 80.

Lasalle A, Guizard C, Maire E, Adrien J, Deville S (2012). Particle redistribution and structural defect development during ice templating. Acta Materialia 60, 4594-4603.

Lösche K, Huen J, Lochte K, Dieckmann G, Bayer-Giraldi M (2014). Einsatz eines neuartigen Antifreeze-Proteins aus marinen Ressourcen (Kieselalgen) in tiefgekühlten Teiglingen, Abschlussbericht AiF 17181 N, FEI Bonn.

Marangoni AG (2004). Fat crystal networks. CRC Press.

Martino MN, Otero L, Sanz PD, Zaritzky NE (1998). Size and location of ice crystals in pork frozen by high-pressure-assisted freezing as compared to classical methods. Meat Science, 50, 303-313.

Otero L, Martino M, Zaritzky N, Solas M, Sanz PD (2000). Preservation of microstructure in peach and mango during high-pressure-shift freezing. Journal of Food Science, 65, 466-470.

Roeder I, Bayer-Giraldi M, Weikusat C, Huen J (2017). Influence of ISP from a polar sea-ice microalga on the microstructure of frozen cream as measured by cryo-Raman microscopy (Poster). 3rd Ice Binding Conference, Rehovot (Israel), hdl:10013/epic.51507.

Stojceska V, Butler F (2012). Investigation of reported correlation coefficients between rheological properties of the wheat bread doughs and baking performance of the corresponding wheat flours. Trends in Food Science and Technology 24, 13–18.

Sun DW, Li B (2003). Microstructural change of potato tissues frozen by ultrasound-assisted immersion freezing. Journal of Food Engineering, 57, 337-345.

Toporski J, Dieing T, Hollricher O (2018). Confocal Raman Microscopy. Springer, Heidelberg.

Van Bockstaele F, De Leyn I, Eeckhout M, Dewettinck K (2008). Rheological properties of wheat flour dough and the relationship with bread volume. I. Creep-recovery measurements. Cereal Chemistry 85, 753-761.

Wildmoser H, Scheiwiller J, Windhab EJ (2004). Impact of disperse microstructure on rheology and quality aspects of ice cream. LWT-Food Science and Technology 37, 881-891.

Zhong Q, Chen H, Zhang Y, Pan K, Wang W (2015). Delivery systems for food applications. In: Sabliov C, Chen H, Yada R, Nanotechnology and functional foods: effective delivery of bioactive ingredients, John Wiley & Sons, 91-111.

Summary

Freezing of bread dough allows considerable gains in operational flexibility in the bakery supply chain, yet at the expense of product quality. Cryoprotectants and cryostabilisers are commonly used to overcome these difficulties, but the market demand for "clean label" products calls for new solutions.

The physical and chemical transformations occurring during freezing, frozen storage and thawing of bread dough are only partially understood. Therefore, the first objective of this thesis was to develop a method based on confocal Raman microscopy, capable of imaging the microstructure of frozen bread dough. The investigations conducted showed that the specific Raman spectra of the main components of frozen bread dough, i.e. ice, liquid water, starch, gluten and yeast cells may be used for assessing their respective spatial distribution in the samples at the microscopic level, enabling the reconstruction of microstructure images. On these images, ice appeared as a continuous three-dimensional network rather than as single crystals. In addition, it appeared that the formation of large ice blocks during freezing is strongly limited by the presence of the other dough components.

Wheat flour has a complex chemical composition, resulting from genetic, environmental and process factors. This leads to the assumption that some flour compositions may be more appropriate than others for frozen processing. Consequently, the second objective of this thesis was to identify, among the analytical methods usually used within the wheat supply chain, those most appropriate to give indications on the suitability of flours for deep-cooling purposes. Methods characterizing gluten functionality as well as water absorption were shown to be particularly relevant.

It is concluded that cryo Raman microscopy is appropriate for investigating the microstructure of frozen dough in relation to process conditions and product quality, with the main advantages being the absence of need for chemical markers and the ability to distinguish between liquid and frozen water. Furthermore, it is proposed that flours appropriate for freezing processes may be specified by bakery companies and developed by the stakeholders of the supply chain, reducing the need for corresponding additives.

Zusammenfassung

Das Gefrieren von Brotteig ermöglicht eine beträchtliche Steigerung der betrieblichen Flexibilität in der Backbranche, jedoch erfolgt dies auf Kosten der Produktqualität. Spezifische Zusatzstoffe werden verwendet, um diese Schwierigkeiten zu überwinden, aber die Forderung des Marktes nach Clean-Label-Produkten ruft nach neuen Lösungen.

Die während des Einfrierens, der TK-Lagerung und des Auftauens stattfindenden physikalischen und chemischen Veränderungen sind bisher nur zum Teil verstanden. Daher war das erste Ziel dieser Arbeit, eine auf konfokaler Raman-Mikroskopie basierende Methode zu entwickeln, die geeignet sein sollte, die Mikrostruktur gefrorenen Brotteigs darzustellen. Die durchgeführten Untersuchungen zeigten, dass die spezifischen Raman-Spektren der Hautbestandteile gefrorenen Teigs, nämlich Eis, flüssiges Wasser, Stärke, Gluten und Hefezellen genutzt werden können, um ihre jeweilige räumliche Verteilung in den Proben auf mikroskopischer Ebene zu ermitteln, was die Rekonstruktion von Bildern der Mikrostruktur ermöglicht. Auf diesen Bildern stellte sich Eis eher als ein kontinuierliches drei-dimensionales Netzwerk denn als einzelne, voneinander getrennte Kristalle dar. Außerdem zeigte sich, dass die Bildung größerer Eisblöcke während des Gefrierens stark von der Anwesenheit der anderen Teigbestandteile limitiert wird.

Weizenmehl hat eine komplexe chemische Zusammensetzung, als Ergebnis zahlreicher genetischer sowie Umwelt- und Prozessfaktoren. Dies führt zu der Annahme, dass bestimmte Zusammensetzungen besser geeignet sein können als andere für die Durchführung von Gefrierprozessen. Infolgedessen bestand das zweite Ziel dieser Arbeit darin, unter den analytischen Methoden, die gängiger Weise in der Wertschöpfungskette verwendet werden diejenigen zu identifizieren, die die besten Aussagen zur Eignung der Mehle für TK-Anwendungen geben. Methoden, die die Funktionalität des Glutens sowie die Wasseraufnahme charakterisieren, zeigten sich hier als besonders relevant.

Daraus wurde abgeleitet, das Cryo-Raman-Mikroskopie geeignet dafür ist, die Mikrostruktur gefrorenen Brotteigs in Relation zu den Prozessbedingungen und zur Brotqualität zu untersuchen. Die wichtigsten Vorteile sind dabei die Abwesenheit der

Notwendigkeit chemischer Marker sowie die Fähigkeit, zwischen gefrorenem und flüssigem Wasser zu unterscheiden. Außerdem wird vorgeschlagen, dass Mehle für TK-Prozesse von den Anwendern spezifiziert und von den vorgelagerten Mitgliedern der Wertschöpfungskette gezielt entwickelt werden könnten, um den Bedarf an den bisher verwendeten Zusatzstoffen zu reduzieren.

Acknowledgements

My deep gratitude goes to Prof. Andreas Schieber, who supervised this work, as well as to his team at IEL Bonn. I am also very grateful to Prof. Peter Köhler who accepted to act as the second reviewer.

I would like to thank Werner Mlodzianowski, Prof. Klaus Lösche and Markus von Bargen, who gave me the opportunity to conduct this work and have supported me continuously throughout my time at ttz Bremerhaven.

A special thanks to my colleagues Iris Auffahrt, Christian Creutz, Imke Matullat, Linda Böhm, Julia Börsmann and Florian Stukenborg for the excellent work atmosphere, the inspiring discussions, the good collaboration in daily work and their support to this thesis.

I am particularly grateful to Maddalena Bayer, Christian Weikusat and Prof. Ilka Weikusat from the Alfred-Wegener Institute who gave me the opportunity to work on the cryo Raman microscope and exchanged with me on ice properties, Mareile Heitmann, Emanuele Zannini and Prof. Elke A. Arendt from the University College Cork for the good collaboration on flour characterization and Susanne Döring from AIBI who supported the project FLOURplus indefatigably.

Finally, I would like to express my love to my wife Elsa, my children Camille, Nils and Noé and my parents Gabrielle and Marcel.

www.ingramcontent.com/pod-product-compliance
Lightning Source LLC
Chambersburg PA
CBHW061300220326
41599CB00028B/5716